P9-DVU-132

WILDTRACK

WILDTRACK

Reminiscences of a Nature Detective

Hugh Falkus

HOLT, RINEHART AND WINSTON
NEW YORK

First published in the United States of America in 1979
by Holt, Rinehart and Winston, 383 Madison Ave., New York, N.Y. 10017.

Library of Congress Cataloging in Publication Data

Falkus, Hugh.
Wildtrack
1. Zoology—England—Lake District. 2. Animals, Habits and behavior of. I. Title.
QL256.F34 1979 591.9′427′8 78-16537
ISBN 0-03-046506-0

Printed in Great Britain

10 9 8 7 6 5 4 3 2 1

For the Home Team

Acknowledgements

To the many friends who have provided help and inspiration for this book, I should like to offer my thanks; in particular to Professor Niko Tinbergen and Dr Hans Kruuk—nature detectives extraordinary.

For his splendid animal pictures, which have never before appeared in print, it is with pleasure that I acknowledge the contribution of Michael Wilkes of Redditch. And special thanks are due to Dr Ron Summers of the University of Capetown for his striking photographs taken on the Ythan Estuary.

I am particularly indebted to Chief Forester, Bill Grant, for his help in arranging various visits to Grizedale Forest, and for the co-operation of John Cubby, Head Ranger, who so ably demonstrated the skills of a professional nature detective.

Major Jimmy Rose, warden of the Ravenglass Nature Reserve, and Dr Brian Spencer gave me much valuable help. So, too, did those well-known 'trusties', Tom Rawling and Fred Buller, who desisted from fishing my river for long enough to use their cameras.

Finally, a special word of thanks to my old friend, Arthur Oglesby, for his superb pictures, which capture so precisely the splendour of the Cumbrian forest and fellsides near my home.

Hugh Falkus
Cragg Cottage
1978

Introduction

Nature detective work, the reading of animal tracks and signs, is a skill that must have been second nature to early man, but (except by the hunter) is now almost entirely forgotten. And yet, for any person of any age who has any interest whatever in wildlife, tracking can become an absorbing pastime.

As shown by the popularity of crosswords or conundrums, there is something in the human condition that responds to the challenge of a puzzle. We relish the intellectual stimulus provided by chess problems, acrostics, jig-saws and other brain-teasers. We identify with the detective, both of fiction and fact, and indeed tend to carry this identity into our everyday lives. Few of us, I fancy, have not at some time paused to consider the characteristics of an unseen stranger who has left footprints in the seaside sand. Somewhere deep inside us a Miss Marples or a Sherlock Holmes is for ever trying to get out.

Human footprints are, of course, immediately recognizable as such. But man is not the only creature to leave a trail of prints behind him in soft ground. Although some of their signs and sounds and scents may be unfamiliar to many of us, all animals leave clues which, if we use our brains, enable us to identify the creatures that have been there before us and to deduce what they have been up to at the time.

Many species of nocturnal animals are difficult to observe in the wild state since most of their hunting is done under cover of darkness. And even when we watch some of the animals that are active by day, their behaviour is not always easy to grasp—for much of it happens so quickly. But in order to discover something about an animal's hunting behaviour we do not *have* to be present while it is hunting. Tracking, whenever conditions are suitable, gives us the opportunity not only to recognize the species of animal that made the tracks, but to build up a picture of its activities. This picture will be just as clear, just as accurate, as it would be if we had been present and watched the animal in action. Indeed, sometimes it may even be more reliable, for the eye can play us tricks.

Tracking tells us the truth about animals.

Five eye-witnesses may give five widely differing accounts of the same incident. But if the tracks of, say, a fox that has hunted and killed a rabbit, remain clearly printed in a stretch of sand, the result is unequivocal. The incident is preserved in its exactitude—until such time as wind or rain blots it out. If we photograph it we have a piece of incontrovertible documentary evidence. There are the fox's prints. There are the rabbit's prints. There is the scuffle; the tufts of fur; the marks of the rabbit's dragging legs as its body was carried away. The story unfolds in front of our eyes as clearly as if we had just seen it happen.

Almost everywhere in the countryside there are stories to be read—if we have eyes to see, and use our intelligence to interpret what we see. Moorland and mountainside, common land, fells, sand dunes, sea shore and estuary, fields and hedgerows, lakeside, river and stream, even the garden at home, offer unlimited opportunities for nature detective work. There are tell-tale signs on plants, bushes, rocks, branches and the bark of trees. In wet grass, sand, mud, snow or soft earth are footprints, scrapes, scratches and scuffles. Even where the ground is too hard to register an imprint, there may be feathers, droppings, pellets, wisps of fur: clues from which we can deduce what creatures have been there before us; what they have been up to, and why.

And nature detective work does not stop at things inanimate.

In the springtime when we see an elver wriggling steadily upriver from the sea on the last stage of its astonishing journey, we realize we are glimpsing one of nature's marvels, its fantastic story uncovered only sixty years ago by Johannes Schmidt's classic piece of detection. Or when on some lonely shore a mewing gull leads its partner in a little jig across the sands towards its nest site in the marram grass, we are watching (if we use our brains to interpret what we see and hear) an example of animal communication as important and as complex as our own.

All my life I have been interested in the natural world about me, both professionally as a writer and director of wildlife films for television, and privately as a sportsman and painter. And over the years I have learned from my own observations and from contact with such superb naturalists as Niko Tinbergen, George Dunnet, Hans Kruuk, Peter Parks, Sean Morris and many others with whom I have had the privilege of working, the fascination of studying the behaviour of animals in their natural habitat, and the value of nature detective work in the understanding of that behaviour.

This is a very personal book, largely about my own neighbourhood and some of the creatures that share it with me. It is intended *not* for the expert, but for the very ordinary person like myself who has no specialized knowledge: an attempt quite simply to pass on some idea of how much extra pleasure and interest nature detective work can add to one's casual enjoyment of the countryside.

No book can hope to deal with every location and animal that might be encountered. Nor does it need to. The principles of nature detection are the same the world over. We do not have to visit the ends of the earth to find something fascinating in nature: it is waiting for us just outside our homes, in the behaviour of those common or garden animals that most of us seem to know so well—and yet know so very little about!

And animals, of course, are not alone in offering us countryside conundrums. As well as pondering the origin of a feather floating on the water, or a twist of hair lodged among the grass stems, or why some animals are brilliantly coloured and some are drab, we may wonder why all the trees of a hilltop spinney are leaning in the same direction, or why one particular piece of ground holds clumps of rhododendrons—while others do not.

The stories we read are sometimes amusing, sometimes dramatic; but always they lead towards a better understanding of the countryside or of animal behaviour. Whatever our interest in wildlife—whether as countrymen of the open air, or town-dwellers whose active interest in nature is confined to holidays and weekends—the reading of these stories, the piecing together of the clues that lead up to them, is deeply rewarding. It adds another dimension to the time spent out of doors.

Above all else, it is fun. And anyone can do it. It requires no special equipment—only an awareness.

But we must learn to use our eyes in a new way—or rather, in a very *old* way: the way of our ancestral hunters. By giving ourselves the chance to flex those hidden hunting instincts which, like wizened muscles, have been inhibited by centuries of urban life, we begin to sharpen our powers of observation and deduction. From this growing consciousness of what is really happening in nature, we start to see and to understand a host of things we have never before even noticed.

And then, suddenly, like Alice through her looking-glass, we find we have stepped into a strange and wonderful new world where nothing will ever be quite the same again.

Part 1

The Coast

The first part of this book deals with tracking in the sand and mud of coastal dunes, estuaries and seashore, many examples of which are taken from the beautiful sweep of coastline a few miles from my cottage in Cumbria.

There are, of course, a great number of similar places—not only in Great Britain but all over the world. I have chosen to describe this particular location because it is rich in wildlife and well documented. Having known it for many years, I am able to discuss at first hand how some of the migrant and resident animals it holds interact; and how, through simple detective work, their behaviour may be viewed with fresh understanding and delight.

Sand dunes at Ravenglass

The Ravenglass nature reserve is set among the wild marram-covered sand dunes of a Cumbrian peninsula. Westwards lies the Irish Sea. The River Irt with its background of fells curves like a crooked finger along the reserve's eastern flank, while to the south the dunes are bounded by an estuary where the rivers Irt, Mite and Esk flow seawards through a single channel.

The peninsula is very old. Its layer of sand rests on a gravel bed deposited by rivers which in the Ice Age must have been torrential. Although Neolithic flint sites in the northern half give evidence of its long existence, there are indications that the sea has often washed over its entire length.

Sand dunes and estuary at Ravenglass. Here for many years Niko Tinbergen F.R.S., Nobel Prize winner and Oxford University Professor of Animal Behaviour, camped during each breeding season to guide his zoology students in their post-graduate studies of animals in the wild. In research of this kind nature detective work is of great value.

The dunes themselves have been formed by sand blown from the sea on westerly gales. Various plants have consolidated this sand and capture more as it is blown in. Sand that drifts eastwards into the River Irt is swept in a right angle out to sea again, whence it completes its cycle by being washed up on to the beach and, when dry, is once again blown landwards.

At any time of the year, like most unspoiled dunes, this stretch of Cumbrian coastline is a place of beauty and fascination. But in the gathering light of a late February afternoon it presents a scene of winter loneliness. Those miles of windswept sand and marram are inhabited by few species other than skylarks, rabbits, sheep and the occasional fox. Earlier in the day it is possible that a flock of Bewick's swans, migrating eastwards from Ireland, may have come trumpeting down to the estuary sands to rest for a few hours before continuing the long flight towards their Siberian breeding grounds. Now, however, the mudflats hold only a scattering of waders common to most estuaries. The swans have gone, perhaps never to return, and there is no whisper of their music in the dusk.

A few herring gulls scavenge along the tideline. A skylark-hunting merlin darts between the steeper sand dunes and flickers across the estuary flats. High overhead some wigeon packs whistle in towards the saltings. A solitary oystercatcher comes piping from the mussel beds. Little else is moving. There are few sounds, except for the wind and the distant sea and the curlews' crying.

But what a fantastic change is coming very soon. During spring and summer, Ravenglass is the breeding ground of one of the largest black-headed gull colonies in Europe. In only a few weeks' time the southern end of these silent and deserted dunes will erupt into a scene of frantic activity. It is an amazing transformation.

Early in March the flock of gulls begins to build up out on the estuary sands. Day by day the numbers swell until, on some suitable morning during the third week in March, the assembled birds rise in a vast white cloud, circle for the first time low over the sandhills and then sweep down together on to their nesting grounds. Suddenly, the dunes are speckled white and the air is loud with relentless clamour.

This first visit lasts only about a couple of hours. Although the birds return for lengthening periods every day, they continue to roost out on the shore until their eggs are laid. By doing this they tend to escape the fox which, from now on, becomes more interested in the gull colony as a source of food. The open sands are safe enough—except on dark, stormy nights, when a fox can sometimes sneak up on the roosting birds and make a mass killing.

'A solitary oystercatcher comes piping from the mussel beds.'

The gulls are followed by other species: Sandwich tern, Common tern, Arctic tern, and the very rare and beautiful Little tern that always nests on the shingle just above high water mark—sometimes within a few yards of ringed plover and oystercatcher. And into the dunes come the mergansers, to lay their eggs in dense clumps of marram grass. Shelducks and wheatears take possession of old rabbit burrows. Adders, voles, lizards and beetles are all present; and very soon, as the laying season reaches its peak, the sands become criss-crossed with the tracks of egg-hunting hedgehogs and foxes.

Natterjacks, too, are active. The shallow pools among the low-lying grass hollows are full of spawn, which lies spread out on the bottom like strings of beads in an endless variety of weird patterns. The Ravenglass Reserve is one of the very few places in

The natterjack toad shown has been photographed leaving its characteristic prints. The nature detective may find the dunes criss-crossed with natterjack tracks, but never catch a glimpse of the animal itself. The reason is that the natterjacks roam about at night hunting for insects, but dig themselves into the sand at daybreak or burrow underneath pieces of driftwood. This prevents them from being dessicated by the hot summer sun.

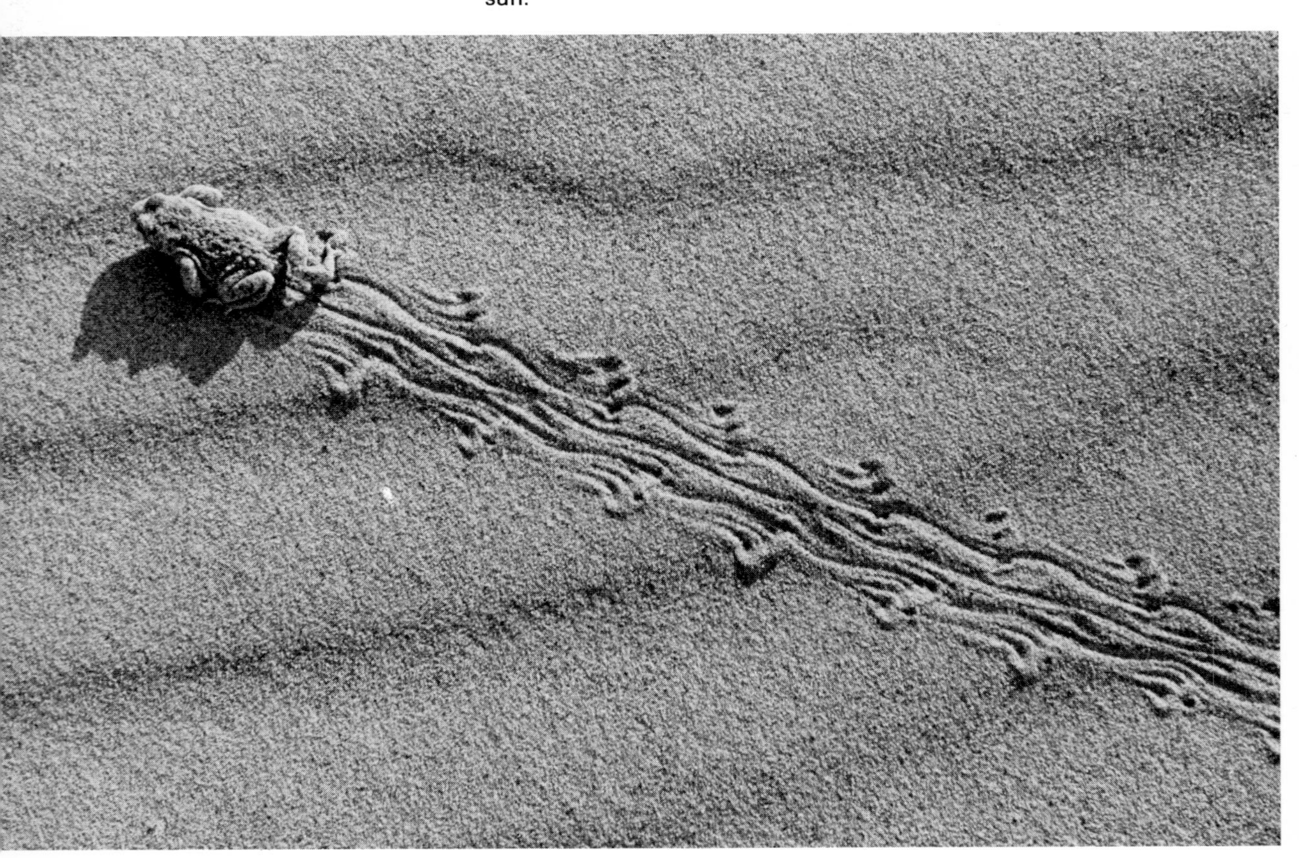

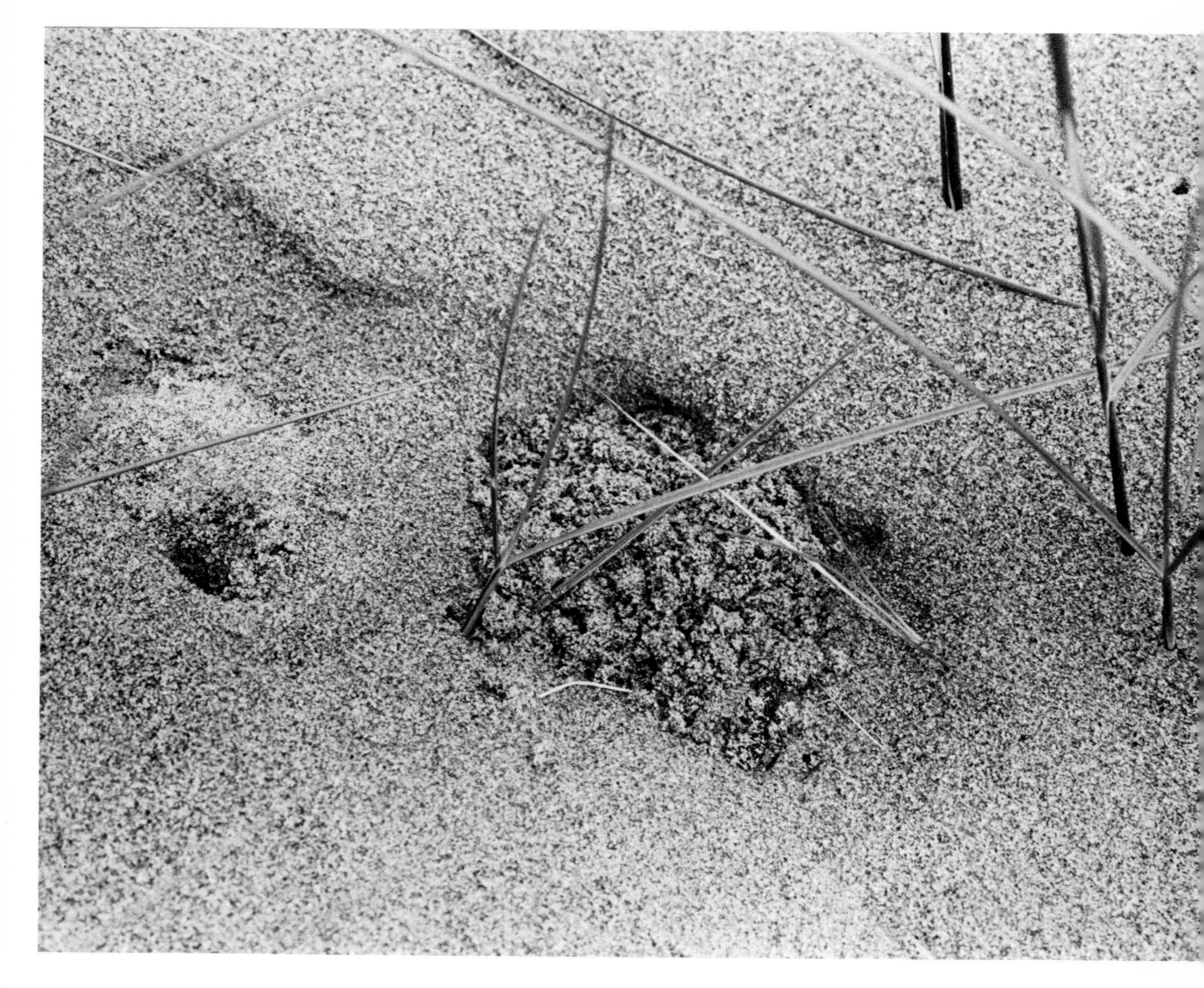

Daytime hiding place of a natterjack toad.

Britain where the rare and beautiful little natterjack is still found.

Sand dunes seem a strange environment for a toad. But the natterjack survives because it is active only at night. During the heat of the day it avoids being dried up by the sun by digging itself into the damp sand, or hiding away underneath some piece of driftwood. After dark it emerges and roams all over the dunes in search of insects, digging itself in again at dawn.

It is smaller than the common toad—on average about the size of a matchbox—and has a distinctive yellow stripe down the middle of its back.

The male's musical call, which it uses to attract a female, can frequently be heard at dusk. The spawning pools are very shallow and in some seasons dry up altogether. When this happens, the natterjacks just skip a year. Often, the pools are there when the

'Like strings of beads'. Natterjack spawn lying on the bottom of a shallow pool among the Ravenglass sand dunes.

natterjacks spawn but dry up before the tadpoles have grown legs and lost their tails—and that's a bad year for natterjacks. During recent years, however, the Reserve Warden has deepened some of the pools, preventing complete catastrophe.

The spawn takes between seven and ten days to hatch into tadpoles, depending on the temperature. In the tadpole stage, lasting eight to ten weeks, natterjacks are very vulnerable and the population fluctuates considerably. In wet springs, tens of thousands survive and the valleys are swarming with little toads that race about at night all over the dunes.

Since natterjacks come out only at night, they are seldom seen in action. But during a season when the early summer has been wet their tracks are plentiful enough. Unlike the common toad, they move very quickly—scuttling along with surprising speed for great distances in search of food.

Natterjack crossroads.

From the third week in March to the third week in April the gulls are busy establishing their nesting territories within the colony. It is a period of intense activity, and very noisy.

Like herring gulls and lesser black-backs, black-headed gulls nest in social groups or colonies. But within a colony they keep their distance from each other. There is a spendid view of the Ravenglass gullery from the top of a nearby dune, and it is striking to see how evenly the pairs of gulls are spaced out, a yard or two apart. This occupation of the ground seems to be a compromise between two dangers: if the nests are set too closely together, neighbouring birds are liable to fight and to prey on each other's eggs and chicks. If the nests are too isolated they are vulnerable to outside predators such as stoat, herring gull or carrion crow, each of which may be mobbed and driven away from a densely nesting colony. If an egg-hunting crow attacks, the birds will rise together in strength and drive it away, whereas the crow could easily rob an isolated nest. *Single* black-headed gulls have no effect at all on crows.

Nesting right in the centre of the gull colony is a small colony of sandwich terns. They make no attempt to defend themselves against predators, seeming to rely entirely on the gulls for their protection.

The distribution of nest sites within the gull colony is governed by a certain amount of fighting; by threat calls, and by postures of intimidation. Many species of animals have 'hostility' gestures that they use against their own species, and are construed as such by other members of the species; they are part of a language that is understood by all members of that group, but not necessarily by members of another group. We instantly understand the significance of a clenched fist, or a shake of the head—but these gestures might well be puzzling to a visitor from outer space.

So it is with the black-headed gull. The gulls are *called* 'black'-headed, but in reality their heads are coloured dark brown. For most of the year their heads are white; the chocolate colouration, which forms a facial mask, occurs only during the breeding season. Experiments carried out at Ravenglass suggest that this dark facial mask is of primary importance as a means of intimidation; that other birds are frightened of it. So—having a dark face helps a gull to keep its neighbours at a safe distance. The gulls' aggressive calls have the same effect. They are 'signals' that help the birds to secure and defend their territories.

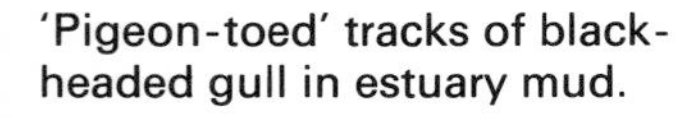

'Pigeon-toed' tracks of black-headed gull in estuary mud.

Once a male has established his territory, and while still busy keeping rivals away, he calls at every passing female. Every now and then one of these females will land nearby, and both birds show off their courtship displays. Often she is afraid of him, and flies off to try her luck elsewhere. But sooner or later one female will keep coming back. Gradually, the two birds become accustomed to each other, and eventually the male feeds his love by regurgitating a bunch of half-digested worms—which she greedily gobbles up. They are now 'engaged' and settle together in the male's established territory.

Then comes the business of building a nest. First of all, the male bird will walk up to a clump of marram grass, bend down and start to coo. This has an irresistible attraction for the female, who goes to join him. Then in turn they sit down and scrape out a cup in the soil with their feet. Both birds collect grass and straws and work them into the rim of the cup to form a nest.

They like to sit in the nest even before they have an egg. Sometimes the female does not fancy the house site that her mate has chosen. In this case she does her best to lure him away to another site. She may succeed. She may not.

Not all the black-headed gulls arrive in the colony at the same time. Some come later and already have a mate. Such pairs go 'house-hunting' together. They can be recognized by the way they fly together in a hesitant, searching way, criss-crossing again and again over the colony. These latecomers often try to land in the densest part of the colony. Invariably, this meets with hostility from the birds already in possession of that territory, and the newcomers move on to try elsewhere. Eventually they find a place where they are tolerated, and in this way all the vacant sites in the colony are filled.

At daybreak

The long, slanting light of early morning is ideal for tracking. Animal prints show up best during that magic hour following sunrise when their outlines are crisp and clear. By breakfast time on a hot summer's day the sun has dried the sand so that it begins to crumble and blur the prints, or a wind may have sprung up and blotted them out altogether. Many tracking expeditions are ruined by wind, or perhaps by a heavy shower of overnight rain that has made the sand too wet and hard to hold clear prints. At other times, a particular animal that one had hoped to track is unaccountably absent. Even so, few mornings are completely wasted.

The stars are dimming and the lights in the distant village are going out. The air is still and keen with the fragrance of salt marshland intoxicating and sharp in the lungs. From the far-off gullery comes a faint cacophony of twenty thousand birds. Closer at hand, the raucous cries of herring gulls flighting towards inland fields. Redshanks twist, piping, along the water's edge. Curlews that are not breeding this season and are spending the summer on the estuary, remote from their fellows nesting in the distant fells, are calling as they flight downriver to land on mudbank and salting.

A rosy tint of sunrise touches the fells and the estuary is ablaze with golden light. Ahead, stretching away among clumps of marram, is a great sweep of yellow sand—a freshly printed book, waiting to be read for the first time.

Heading into the dunes: prints of fox and hedgehog (right).

Even after myxomatosis, the most familiar tracks are those of the rabbit. Anyone who has walked across sand dunes or snow-covered fields will be familiar with the path of the rabbit. But even so, a surprising number of people—and not only the newcomer to nature detective work—would find it difficult to answer the following questions:

1. Has the rabbit that has left the prints in this picture been travelling from right to left or from left to right?
2. Which feet have made which prints?

1. The rabbit has been hopping along from right to left.
2. The two fore-and-aft prints at the back were made by the *forelegs*. The two side-by-side prints in front (arrowed) were made by the back legs—swinging forward as the rabbit hopped.

The rabbit (historical)

The rabbit, or coney, has been the subject of much charming literature, notably the Beatrix Potter stories, and Richard Adams's *Watership Down*, but the sentimental aspects of this little animal should not be allowed to obscure its darker side.

The rabbit is a menace. Ever hungry, it will eat almost anything vegetable and in an uncontrolled state does enormous harm to forestry, agricultural crops and wild plants. It is not, as many people think, a native of our countryside, but was introduced to Britain probably in Saxon times. It is mentioned in a charter during King Canute's reign.

For centuries, rabbits were domesticated animals and a valuable source of fresh meat, being kept and bred for food in special enclosures, or 'warrens'. The word *warren*, whose root means 'to protect', persists in many districts, although with the exception of occasional warrens that were built of stone—such as Ditsworthy Warren near Dartmoor—few traces of the original enclosures remain.

When the warrens fell into disuse the rabbit went 'wild', but seems to have been kept in check by a hungry rural population. There is no mention of it as a pest before about a hundred years ago. From then until the early nineteen-fifties its numbers multiplied and become uncontrollable, until—in 1953-4—the virus of myxomatosis disease appeared. This disease spread with

Tracks of rabbit that has been hopping from left to right. The rather curious circle has been described by a stem of marram grass blowing in the wind.

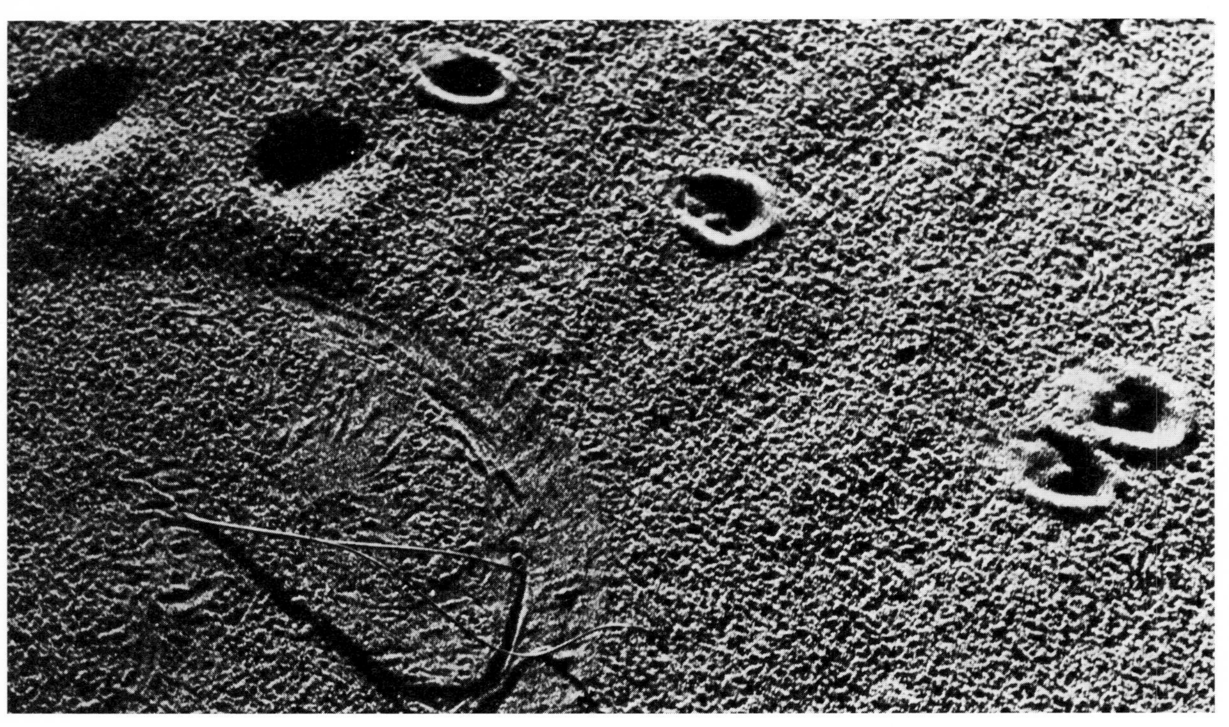

great rapidity to almost every part of Britain and almost wiped the rabbit out.

Almost, but not quite.

Recently, a strain immune to the disease has started to establish itself, and there seems no likelihood that the British rabbit will ever be exterminated.

Signs of a love contest amongst the dunes. Rabbit 'mating rings'.

This small scrape in the soil indicates a feeding scene. Rain has obliterated footprints, but the droppings tell us what animal has been here: a rabbit, digging for roots.

The rabbit (domestic)

In a sandy basin between dunes a line of scuffled prints shows where one rabbit has been chased by another and made to drop its meal—a piece of moss scraped up from a damp hollow.

Particularly noticeable are numerous little holes scratched out by rabbits, the sand in front usually sprinkled with droppings. Not abortive attempts at digging a burrow, as one might think, but food scrapes. Rabbits will eat almost anything that is vegetable.

Rabbit mating 'rings' are found everywhere among the dunes. They are made by two bucks chasing each other round over rivalry for a doe, scattering the sand, biting each other and literally 'making the fur fly'. After the tussle the winning buck claims his lady love, and begins his nuptials by urinating over her—presumably to 'stake his claim': a rather weird aspect of the rabbit's mating ritual.

At a spot along the side of a dune, rabbit tracks converge to form a main 'trod' or pathway that is obviously in regular use. It leads to a sign so slight that many of us would pass it by without a second glance. What seems no more than a casual disturbance in the sand marks the closed entrance to a rabbit 'stop' or nest, the blind burrow dug by a doe shortly before she gives birth to her litter of young. She leaves the young rabbits in the stop for most of the day, returning to feed them usually at dusk and daybreak, digging up the entrance when she arrives and closing it again when she departs.

Rabbit 'stop'.

A rabbit stop (arrowed) set high among the sand dunes. The tracks of the doe can be plainly seen, curling round the top of the dune.

One of the inmates of the stop. Young rabbits can run about at the end of a fortnight. They are self-supporting after four weeks and can start to breed when six months old.

Both foxes and badgers can locate closed rabbit stops by scent, and will kill and eat the young. The badger will frequently dig them out, but the fox seems to raid a stop only when it 'happens' past during a doe's visit, and the entrance is already open. (For pictures of a rabbit stop dug out by badger see p. 81).

Some of the old rabbit holes soon have new tenants when the wheatears and shelducks move into the dunes. There is no doubt about the identity of the occupants when a hole has been taken over as nesting site by a pair of shelducks. Their tracks can be seen leading in and out of the burrow.

Shelduck tracks leading in and out of abandoned rabbit burrow.

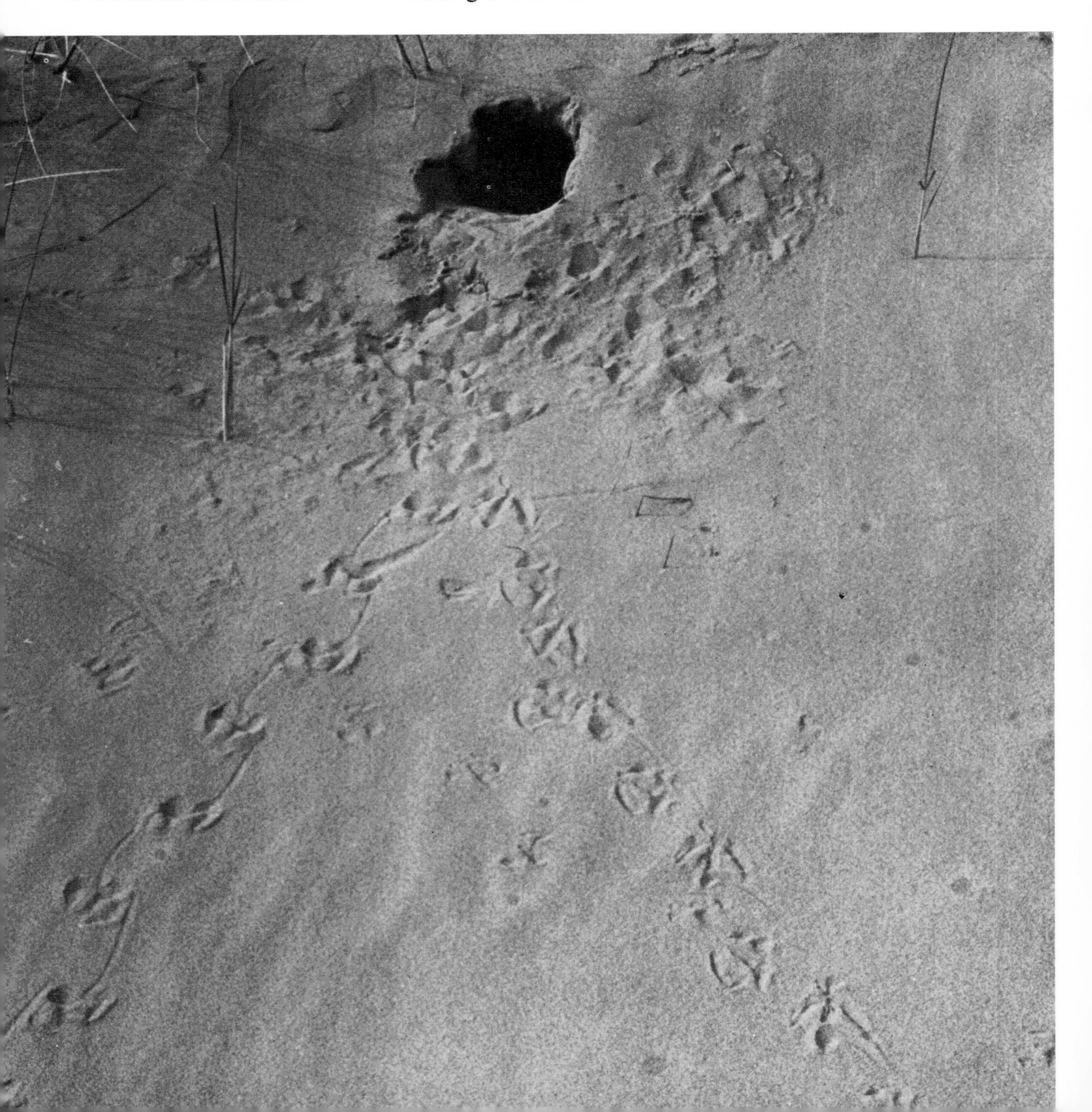

The birds themselves are not far away. Through binoculars from a dune top they can be seen out on the estuary mud; the female dibbling for food, while the male engages in elaborate postures of intimidation to warn off other males.

Later, one may find a set of shallow ribbon-like marks in the mud. These are typical shelduck feeding tracks, made as the birds skim along with their beaks after tiny snails and shrimp-like crustaceans that live in the surface film (see p. 103).

Cinnabar caterpillar moving over sand. Its bold markings have a most important function (see p. 85).

Hare, from Topsell's *Historie of Foure-Footed Beastes*, 1607.

The hare does not seem to have been a popular animal in the British Isles. According to Caesar, the Britons wouldn't eat it—although the Romans themselves regarded it as a delicacy. Indeed, it seems always to have been thought of as a magic beast. Witches were believed to assume the shape of a hare. It was said that if you shot a hare with a silver bullet you would find a witch dead in her cottage with the bullet in the place where the hare had been hit. Again, if an expectant mother stepped over a hare's form (or couching place) her child would be born with a hare lip—a belief common all over the Continent.

It is true that, after giving birth, the doe carries each leveret in her mouth to a separate 'form'—just as a cat carries her kittens. By dispersing them in this fashion she gives them a greater chance of survival. There are between one and four leverets in a litter and they are born fully furred with their eyes open. A marked difference between the hare and the rabbit—whose young are born naked and blind in a burrow. Leverets are very much on their own from birth—except at night, when the doe comes round to suckle them. There are several litters in a season, and leverets have been recorded in every month of the year.

Tracks (or 'prickings') of a brown hare (left). The hare has poor eyesight but very fast legs and depends entirely on its speed and footwork to elude pursuers. Its hind legs are nearly twice as long as its forelegs, and it can run uphill just as fast as it can on the flat—if not faster! Its top speed touches 45 m.p.h.

Opposite
A hunting fox (left) and wandering farm dog have made parallel trails along the shoreline—though certainly not at the same time!

Strangers on the shore, early birds and adders

Tracking on the beach can reveal strange stories. Although we may not often see them during the day, some rather unexpected animals visit the shore at night. An unlikely animal to find far out on the beach, one might think, is the hedgehog. And yet, as their tracks prove, night after night during the summer hedgehogs will come out from the dunes at Ravenglass and roam for extraordinary distances across the open sands. They are known to eat the eggs and chicks of ground-nesting birds, and to hunt for insects along the tide line, but their tracks are frequently found near the *low* tide mark, half a mile from the dunes. There are no signs of feeding activity, and the reason for this wandering is a mystery. What it is that the hedgehogs seek so far out on the sands, nobody knows.

The rabbit, too, is a beachcomber. It often feeds on sea rocket (*Glaux maritima*), a plant that only grows where the seeds have been washed in by sea—the reason why sea rocket grows so close to the high tide line. And so the rabbit becomes a beachcomber by feeding on what the tide has washed ashore.

Story in three parts.
A merganser has landed on the sand (centre top), sat down to have a rest (centre right), then walked on towards its nest site in the marram grass — its tracks running towards bottom of picture from right to left.

Desert tragedy. Common frog that could hop no further. It has been caught out on the open sand after daybreak, unable to reach cover before being dried up by the sun. A fate that overtakes the unwary natterjack.

Like rabbits and hedgehogs, foxes often come out on to the open sands at night. Tracking shows that foxes feed to some extent on carrion; for instance, on sheep that have been drowned, when the river was in flood, and swept out to sea. Subsequently, the carcases are brought in by the waves and stranded on the shore.

Skylark tracks wind everywhere among the clumps of marram. The nests, which are beautifully camouflaged, are in the tussocky grassland that stretches across some of the valleys between the dunes. Although the area of the dunes is very arid, the skylarks are most efficient hunters; their main food being insects and insect larvae, in addition to worms and seeds.

Tracks of greylag goose lead down towards bottom right. At extreme edge of picture the bird has turned round and accelerated towards take-off, leaving imprint of wing-tips in the sand as it became airborne.

Small weevil on dew-damp sand.

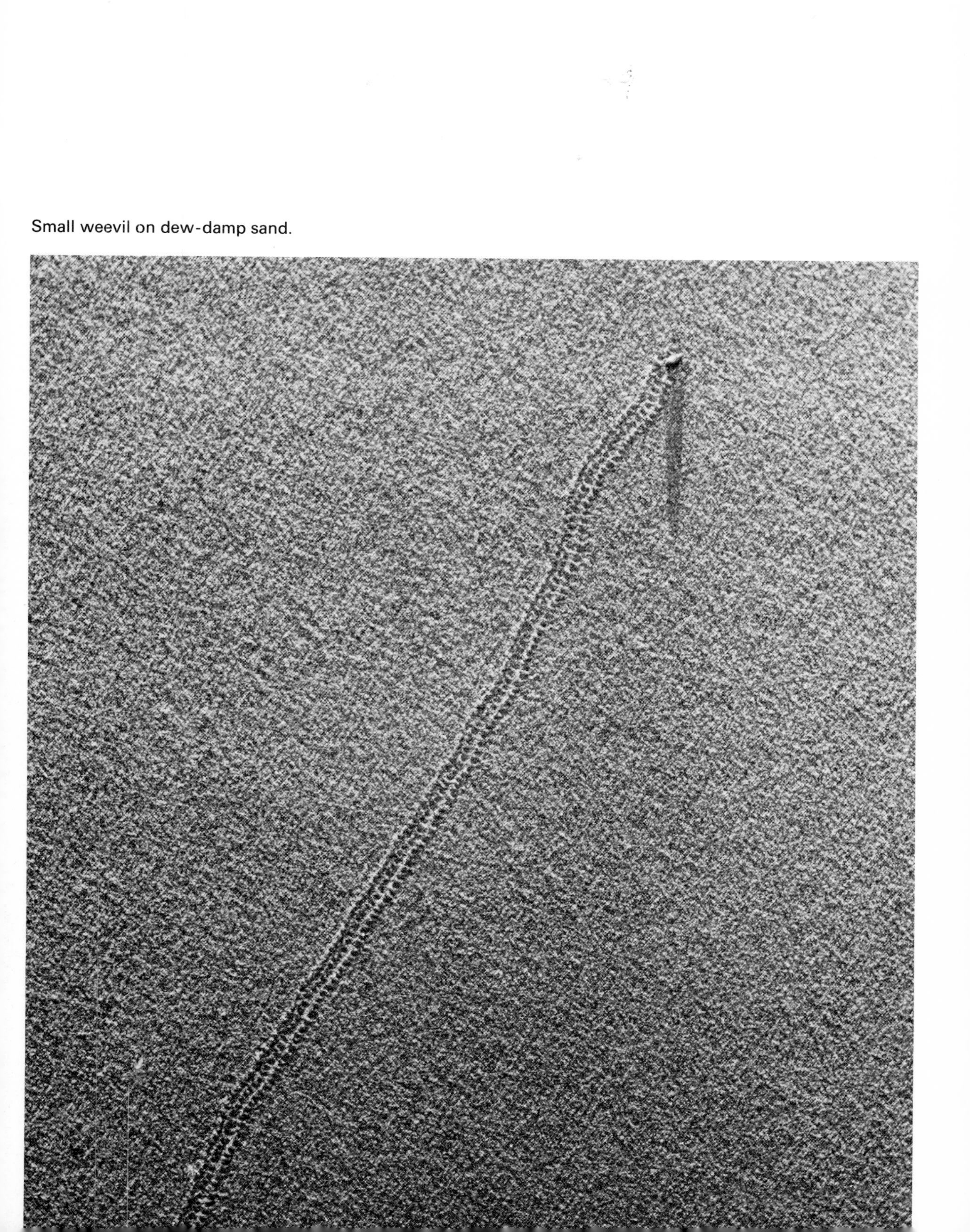

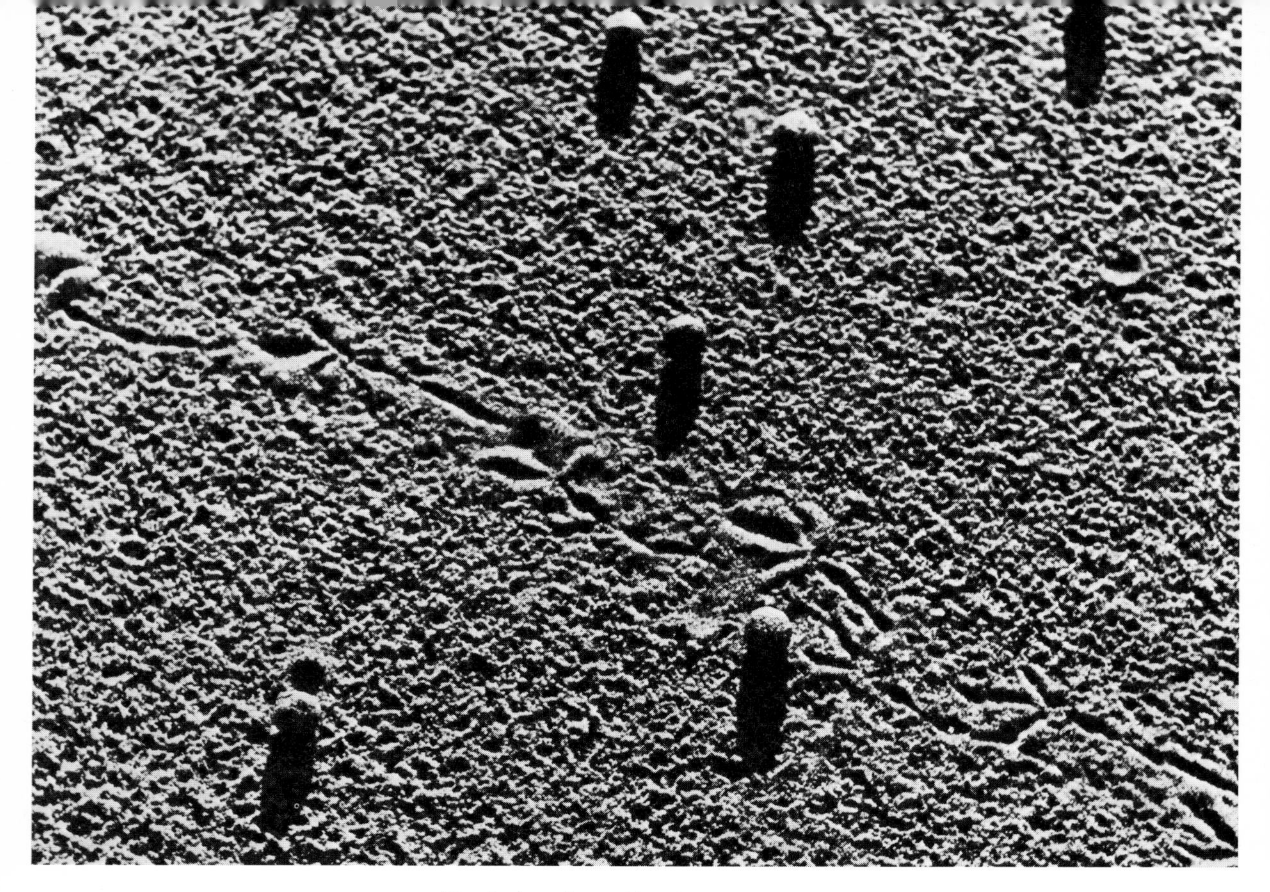

Tracks in a 'lunar' landscape—sand pitted by a slight shower of rain. Skylark, moving from right to left, through rabbit droppings.

Duneland with all its sandy hillocks provides a greater nesting potential to skylarks than a similar area of arable land. This is because the hillocks cut down most of the territorial disputes that would occur if the ground were flatter. When a male bird drops into its nest territory, it loses sight of other skylarks, and is itself hidden from them, behind the surrounding tussocks. This reduces the amount of aggression between males, and permits a greater breeding density.

Skylarks make excellent alarm clocks. Birds get up at different times, and the skylark is one of the earliest. It is not without good reason that we speak of 'rising with the lark'. They usually have two broods, but suffer from hedgehogs and crows—both of which steal eggs and young. Adders, too, will take young skylarks; indeed a single adder can wipe out a complete nest of fledglings. The adder is Britain's only poisonous snake. It is easily distinguished from the harmless grass snake and smooth snake, the only other British species, by the bold zig-zag line that runs down the back. An adder bite is seldom fatal, but it is very painful.

Before you read the caption below. What animal has been here?

An adder. It has left these curious curving tracks as it crawled in 'concertina' fashion across the sand soon after sunrise.

An adder crawling over bare sand moves in an unusual way for a snake. It begins by lying folded up, like a concertina. Next, it stretches the front part of its body forward, using the tail as an anchor. Then it begins to fold the front part of the body, dragging the tail along. This continues until the animal lies all folded up again, in the same position as at the start, only a little further forward.

Adder moving across smooth sand with 'concertina' movement.

Compare markings of adder with those of the grass snake opposite.

On a hot summer's day, adders like to 'lie out' basking in the sunshine. More than once I have narrowly avoided putting a hand on one when leaning forward to peer over a low grassy flood bank by the riverside near my cottage.

Adders hibernate during the winter months, the period of hibernation being roughly between mid-October and mid-March. They will travel anything from a few hundred yards to a mile during the journey from their hibernation areas to their summer habitats. The critical temperature that governs this seems to be 8°C (46°F). If the temperature is above this figure, adders may come out. If below, they will almost certainly stay in.

The largest British snake, the grass snake is completely harmless. Its main diet consists of frogs and the common toad.

Grass snake's eggs in heap of horse manure.

The mating period is between mid-March and mid-May. When two males come together during this season they offer one of nature's most fascinating scenes: the 'dance' of the adders. This, of course, is not really a dance; nor is it a contest to decide territorial boundaries, as some writers have suggested. Adders do not establish territories—unless the space immediately surrounding a female be construed as a territory. The 'dance' is a tournament; a trial of strength in which two males compete for the favours of a female, who is usually present when the contest takes place.

The males make no attempt to injure each other. Weaving and feinting, they grapple like two all-in wrestlers striving for a hold, the end coming when the dominant snake puts coils round his opponent and throws him on to his back. He may do this several times before chasing his defeated rival away. Occasionally, more than two males will compete simultaneously for a female.

The young adders, usually between ten and fourteen in number, are born alive during August or early September.

Adders are deaf, and make no sound except for a soft hiss when annoyed—the result of drawing in and expelling air from the lungs. They hunt by sight and smell; their usual meal times being at daybreak or in the evening. But they will feed at any time when hungry, if the weather is warm.

The adder is a predator. Its diet includes: mice, voles, shrews, small birds and their eggs (often skylarks'), frogs, newts, lizards, slugs, worms and insects.

A weird pattern indeed. Anyone could be excused for failing this test in animal tracking. It is that adder again! This time it has been struggling up a gentle slope on smooth sand.

Adder struggling to find a
purchase on soft sand slope.

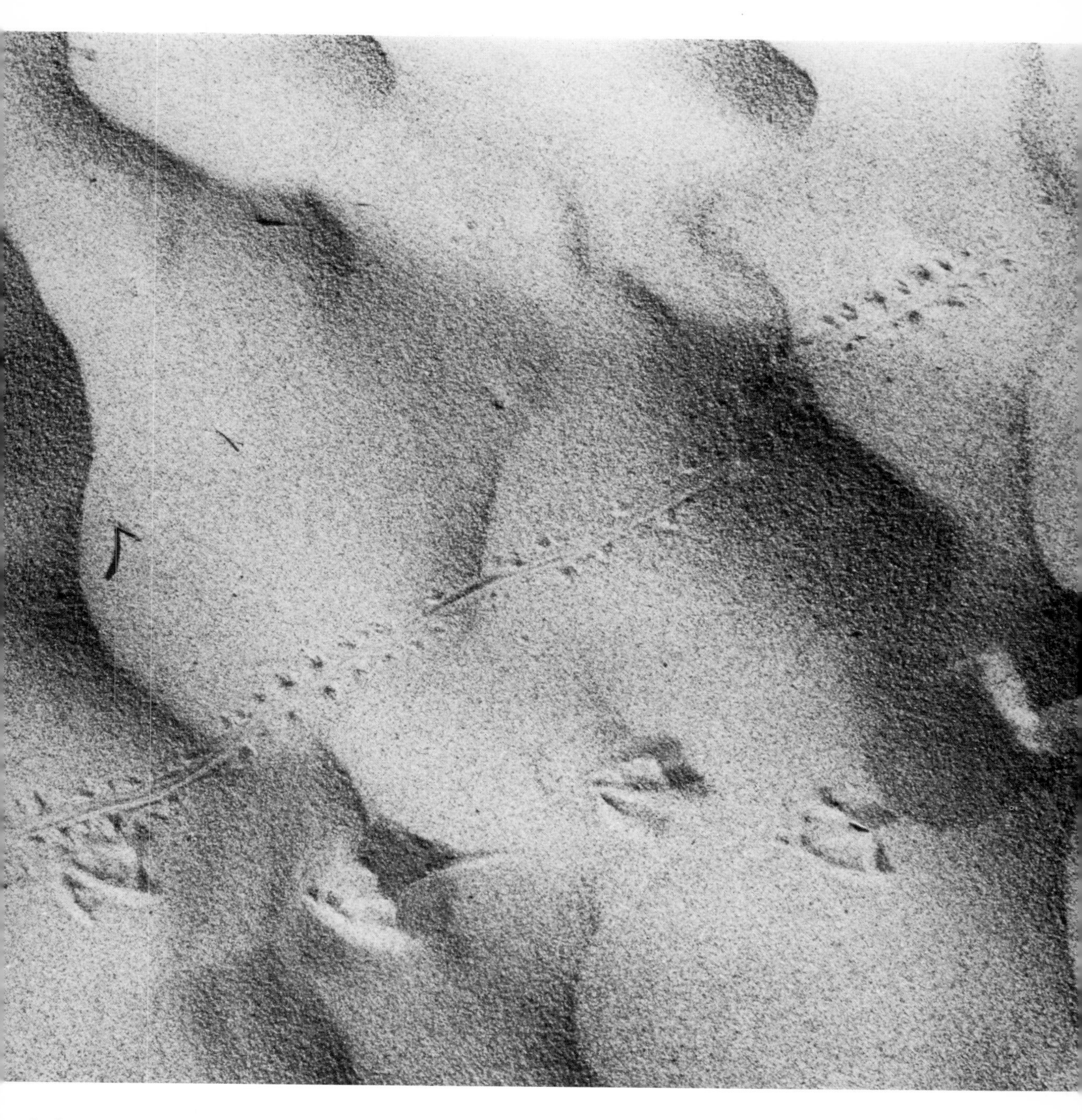

Animal tracks frequently run side by side or criss-cross one upon another. What has happened here?

This story is quite simple and straightforward. A lizard has crossed the path of a black-headed gull. The gull has been walking from left to right, the lizard from right to left—sand can be seen piled up *behind* the foot prints where the animal has been going uphill. The dragging mark of the lizard's tail is very clear.

These are familiar prints. Who made them and in what order?

The bird tracks slanting to the right are the typical dragging footprints of a carrion crow. The centre tracks are those of an adder. Those on the left are from a natterjack toad.
Order of appearance: 1 Crow. 2 Adder. 3 Natterjack.

The crow must have been here before the adder because the adder's tracks cross on top of the crow's. But the natterjack's tracks are on top of the adder's—so the natterjack must have been last of all.

Tracks of ringed plover.

The stoat and the blackbird

The sign of a stoat kill is a small wound on the back of the head. But unless an observer arrives on the scene soon after a killing, corpses are seldom found. In nature nothing goes to waste. The carrion eaters: foxes, rats, crows, ravens, buzzards soon clean up the carcases of dead animals.

The hunting behaviour of a stoat is very interesting to watch. More than once I have seen a stoat hunting rabbits. Each time, it seemed to me, the stoat had marked down, and was intent on pursuing, one particular rabbit; and once it had started to stalk this chosen animal, ignored the others.

One afternoon, however, I witnessed some astonishing behaviour. At least, I had never seen anything like it before, nor have I since.

Coming to a gateway leading into a cow pasture, I noticed a female blackbird pecking at some dried cow pats about fifteen yards from the gate—and thought nothing of it until, just as I was about to open the gate, I saw a stoat hopping across the field. Fortunately hidden by the gatepost, I stood absolutely still. The stoat's course led straight towards the blackbird and I wanted to see what would happen.

It was worth waiting for.

As the stoat approached, the blackbird (to my surprise) gave no note of alarm—continuing to feed as though quite unconcerned.

When within a few yards of the blackbird, the stoat stopped and stood up on its hind legs. The blackbird merely glanced at it and went on pecking. Then, after a pause, the stoat started to circle the bird, quite slowly, but with an amazing repertoire of contortions. 'Acrobatics' would be a better description. It hopped about, leaped up and down and rolled over and over, like an acrobat going round a circus ring in a series of cartwheels, hand springs and somersaults. And as the stoat circled, I noticed that it was edging closer and closer to the blackbird—which was still pecking away with only an occasional sideways glance.

Suddenly, the stoat stopped capering—and pounced.

At the very last moment—almost, it seemed, as the stoat was upon it—the blackbird fluttered straight up to a height of six or seven feet. Then, as the stoat bounded over the cow pats, the bird dropped straight down again and continued its meal. And from its original distance of several yards, the stoat once more started to caper round, circling the blackbird as before, again edging closer and closer as it went round, before pouncing . . . with an exactly similar result.

This pantomime continued for well over twenty minutes, during which the stoat must have made six or seven ineffectual pounces; the blackbird fluttering up each time and coming straight down again.

Eventually, the blackbird seemed to get bored with it all—or perhaps it had simply had enough to eat. At any rate, it flew off into a nearby hawthorn. Whereupon the stoat continued its interrupted journey and disappeared along the hedgerow.

Opposite top
Although some signs are very faint, the story behind them may be highly dramatic.

This picture tells of a killing. (Who is the killer? Who has died?) To the nature detective the story is as clear as if he had seen it happen. The picture caption might be: *How to dispose of a body!*

The trail was left by a stoat that has dragged its prey (a rabbit) across hard sand. The rabbit's body has made the broad line close to the stoat's footprints; the rabbit's dangling legs have left a thin, less pronounced line on the outside.

Opposite bottom
A stoat that has been travelling from right to left has paused briefly to investigate the prints of a young rabbit moving in the same direction. Perhaps this was the very animal the stoat was hunting. There is no doubt that the rabbit was here *before* the stoat, one of the stoat's prints is superimposed on a print made by the rabbit's right hind foot.

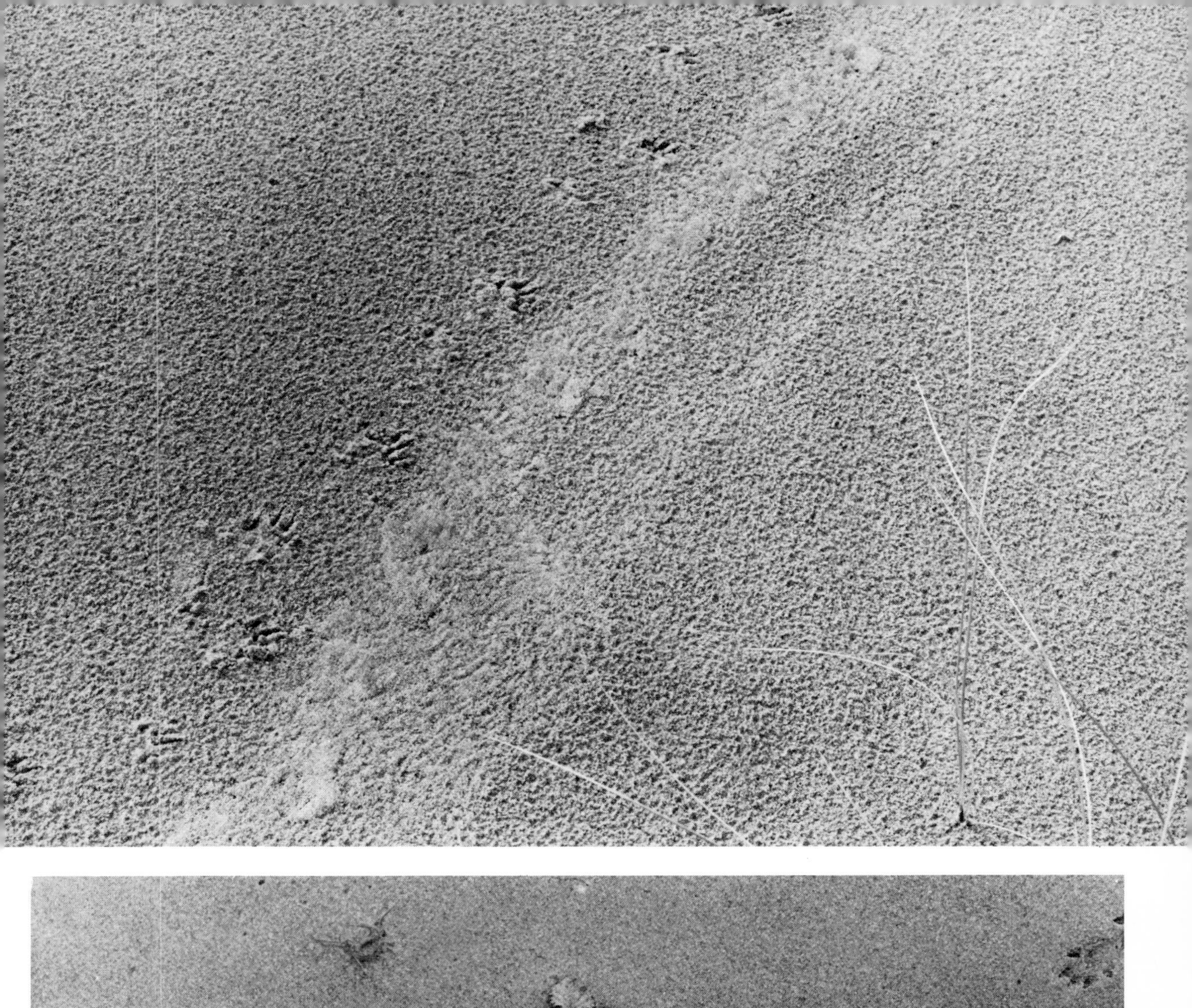

The body of the rabbit has been left almost as a 'husk' among the marram.

Stoat tracks in close-up.

Opposite
The stoat hunts both by night and by day, feeding mainly on rats, mice and voles. It is also a great rabbit hunter, although myxomatosis has reduced its opportunities. Fish and birds are eaten and, in particular, the eggs of ground-nesting species: pheasant, partridge, duck, plover *et al*. An egg is usually eaten on the spot, but is sometimes rolled away. A stoat is said to carry an egg by bracing it between forepaws and chin, and to hop along on its hindlegs—but I have never witnessed this myself.

Opposite
Tracks of a brown rat. The mark left by the tail is very clearly defined.

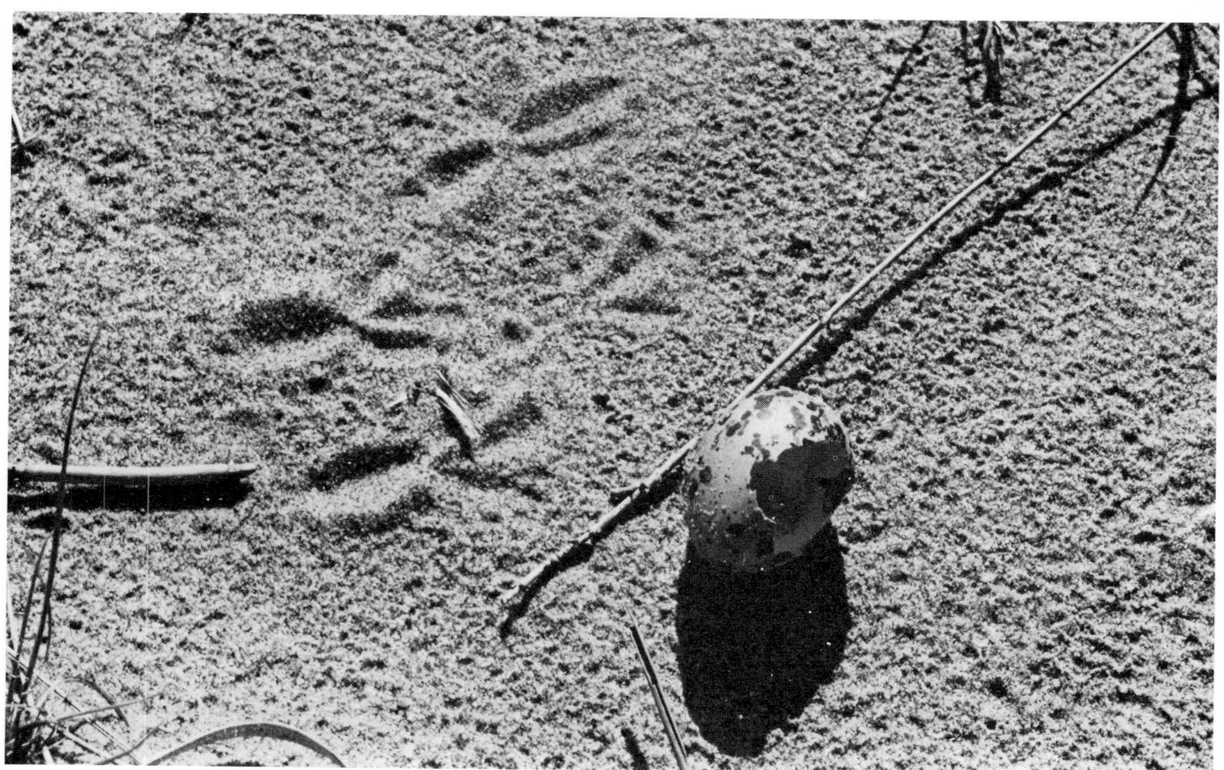

Stop thief!
Badgers, foxes and stoats are all very fond of eggs, and so is another wily predator, the crow. Above is a beautifully camouflaged oyster-catcher's nest with a new clutch. One at least of these young birds is doomed. The picture below shows a stolen egg, and the tracks beside it identify the crow as the culprit.

Fox prints leading to cache where common tern's egg has been buried. The marks of the fox's whiskers can be plainly seen. Skylark tracks wind from left to right. At the top of picture are the tracks of a common tern. The tern has been here later than the fox, and the fox later than the skylark.

Of all the predators that ground-nesting birds have to face, the fox is the most dangerous. The behaviour of black-headed gulls is well adapted to give these birds a reasonable degree of protection against most enemies; but against the fox their defence is inadequate.

A typical fox is about 42 inches from nose to tip of tail, or 'brush' which is about 16 inches long and usually tipped with white. (This white tip, incidentally, is *not* exclusive to the male fox. Both sexes can have it.)

A fox's hearing is very sensitive. It is supposed to be able to hear a watch ticking at 40 feet. But its nose is perhaps its most important sense organ. Being mainly nocturnal, the fox detects its prey, its enemies, and fellow foxes mainly by scenting them.

Fox trot

Egg-laying in the gull colony starts during the third week of April, the peak period being the 25th, and at this stage of the season a complete change takes place. After the raucous sounds of fighting and pair-formation, there is a period of comparative peace and quiet.

But now that the eggs are laid, it is no longer possible for both birds to roost out on the open shore. One bird must always be on the nest, and here it is much more easily attacked by predators, most dangerous of which is the fox.

There is no doubt about the damage a fox can do to ground-nesting birds. The evidence is laid out for the nature detective to read.

Comparatively little is known about the hunting behaviour of foxes, they work mostly by night or in the half-light of dawn, and much of what *is* known has been discovered from reading the signs they leave behind them. In the early morning, before wind and rain wipe out the tracks, the story of the night's activities is printed in the sand.

By the mouth of a fox's den among the outlying dunes is the carcase of a shelduck that has been pulled about and played with by the family of cubs. Near to the den, cub prints show where the young foxes have been playing with a piece of stick, fighting over it and rolling about. And a little distance away are the remains of a rabbit, surrounded by prints made by the cubs when they tore to pieces the food brought to them by their parent.

Further on, a line of fox prints leads towards the distant gullery. Foxes have their regular highways, and here are the parallel tracks of two large, heavy foxes with the lighter and more rounded prints of two cubs.

A family outing. But the cub on the right has been hurrying to catch up with the others!

Fox cub playground.

There are usually five cubs in a litter. As they grow bolder they spend more and more time at play outside the den. But they do so very silently. To attract too much attention when they are so vulnerable would be fatal.

Above
Tracks of fox cubs playing. Crow tracks on left.

Prints of a fox cub. Cub prints are smaller and more rounded than those of the adult, and the claw marks are less pronounced.

Weaning starts when the cubs are about twelve weeks old. From then on, foraging for food becomes all-important.

Above
Prints of Labrador dog. Compare with fox prints (right) seen in parallel with hedgehog prints.

Opposite
The fox has been following the track of a herring gull. Hedgehog track crosses the fox track at top of picture.

Above
Fox and natterjack tracks.

Fox and hedgehog have been going in the same direction, from right to left. But two of the hedgehog's prints are superimposed on a fox print, showing that the fox was here first.

A couple of hundred yards along the trail one fox has struck off on its own, its tracks running alongside those of a hedgehog. Both animals have been heading for a meal in the distant gullery.

Then comes an amusing detail—the fox has suddenly stopped and sat down to scratch itself, leaving a perfect imprint of its tail with even the individual hairs clearly visible.

And here among the marram are the remains of the fox's first kill of the night—a rabbit, whose skin has been turned inside out, like a glove.

The deep footprint (left, foreground) shows where a fox has 'braked' suddenly and squatted on its haunches facing right. It has not been sitting upright, however, since the mark of the left haunch is deeper than the right one, and beyond it is the print of the left buttock. The outline of the brush is clearly visible, with traces of individual hairs, but it is not in line with the middle of the haunches. The fox must have sat leaning over on its left haunch, probably to scratch itself.

Opposite top
Typical fox work: rabbit skin turned inside out.

Opposite bottom
Fox prints are more slender and pear-shaped than those of a dog of similar size. These prints are exceptionally clear because the ground was firm after a light shower of rain. The fox has been going at speed up a steep slope.

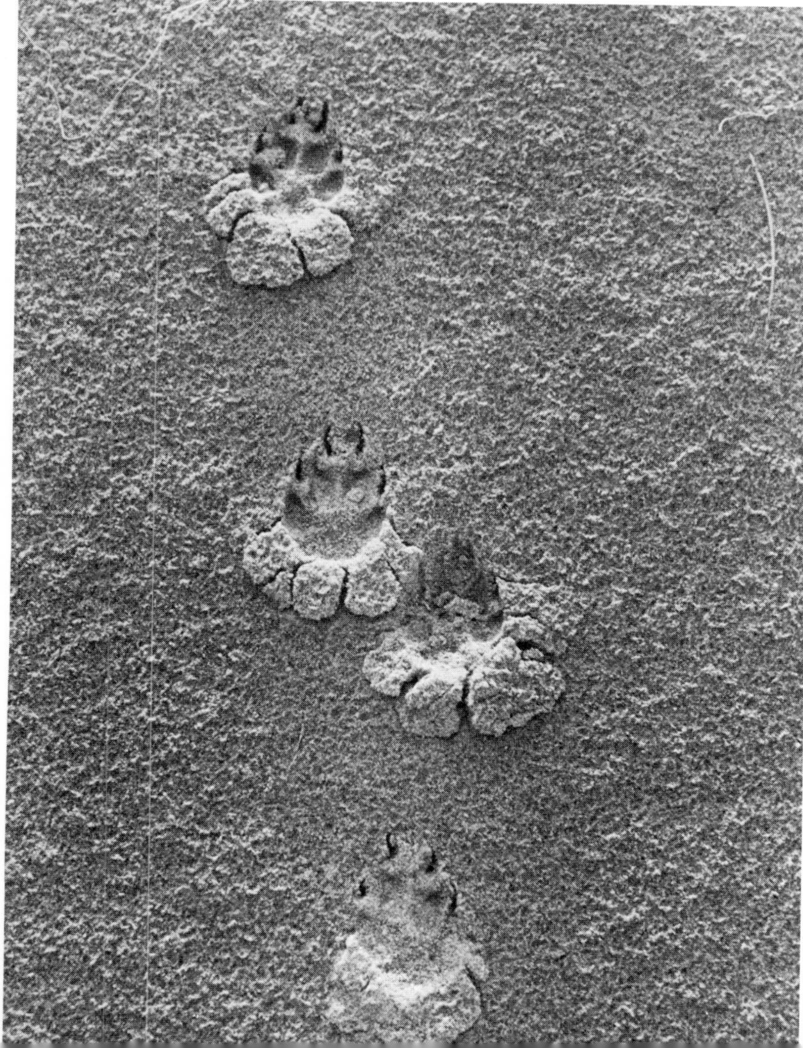

A hundred yards further on, the fox's tracks lead to the remains of a fledgling black-headed gull (pictured below). The long, dragging prints of a crow also surround the kill. Carrion crows are quickly on the scene at daybreak to clean up after a fox has made a killing.

Opposite
Fox slaughter. Black-headed gulls, with sandwich tern (in foreground). Four out of twenty-three birds killed on a dark rainy night while roosting on open sands near gullery. The fox it seems first killed to eat. And then, although its hunger must have been satisfied, has gone on and on killing.

Black-headed gulls are not really birds of the sea shore. They belong to inland fields, where they feed mainly on worms and insect larvae. They are the 'seagulls' so often seen following the plough. Their usual breeding habitat is among the moorland pools and swamps where they are comparatively safe from foxes. Colonies among coastal sand dunes are rare. The gulls' behaviour just isn't adapted to living among dunes; at Ravenglass they are using a very unfavourable location. In one sense it is comparable to a chicken run, where foxes may kill dozens of birds in a brief killing frenzy simply because the hens have no escape route.

On one occasion in the Ravenglass gullery, over two hundred adult birds were killed in a single night.

That the fox should carry out such slaughter seems extraordinary. But surplus killing has a function. Almost all carnivores kill more prey than they need for their immediate requirements, if they get the chance to do so. Opportunities of making a kill are limited—so they must take what chances they can, and cache any surplus food. And the fox is no exception. It by no means stores all the food it kills, but it will hide or bury a good deal of it. (The pictures on p. 65 and p. 67 are examples of this).

Losses caused by fox predation, combined with the damage done by other predators, and by the gulls themselves—by thieving each other's eggs and chicks—seem excessive. And yet sufficient young birds survive to maintain the number of breeding pairs, except for those years when more foxes than usual are working in the dunes.

One might think that the mass laying of eggs, all at the same time, was a weakness in the gulls' behaviour adaptation, and that losses due to predation would be fewer if the laying season were extended over a longer period. But this is unlikely. Social nesting is pointless unless all the birds nest at the same time. If they did not they would be unable to join forces in a combined attack to drive some egg-stealing herring-gull or carrion crow or stoat out of the colony.

Opposite top
Here among the marram grass a fox has dug a hole and buried surplus prey. These scrapes are dug with the forepaws then filled in by shovelling with the snout. In this picture faint snout marks can be seen, bottom left.

On a dark night nesting gulls are easily approached, indeed they may sometimes be lifted from the nest by hand, so that a quietly moving animal like the fox can slaughter almost as it will. The darker the night, the more gulls get killed. Foxes are able to walk quietly from one nest to the next, killing whatever comes in front of them.

Opposite bottom
The cache uncovered. Contents: one young black-headed gull. Many eggs and young birds are hidden by the fox during the breeding season.

Top
Fox kill: black-headed gull.

Bottom
Sandwich tern killed at night by fox.

Sometimes a fox leaves dead birds lying where it kills them; sometimes it carries them a short distance and hides them—either by burying them, or by pushing them into the undergrowth. A gleam of white from a dense patch of stinging nettles may show the nature detective where a dead gull has been tucked away.

Of all the predators the fox is the most destructive, but there is a limit to what even the fox can destroy in a short period. By laying their eggs all at the same time, the gulls reduce this vulnerable period to a minimum. And although the fox is responsible for a great deal of surplus killing and egg stealing, it simply *has* to leave many nests undisturbed. If the laying season were extended, the fox could go on and on killing, eating and burying over a much longer period, and there can be little doubt that the losses in the gullery would be proportionately higher.

By consistent tracking on the dunes morning after morning whenever weather conditions were suitable, Niko Tinbergen established beyond doubt that foxes bury surplus food on a large scale—particularly eggs.

Egg cache surrounded by fox's prints. The slight grooves (arrowed) show where the fox has pushed sand over the buried egg with its nose.

Left
Fox cache opened by hand. Contents: one merganser's egg. Note the prints of the fox (arrowed).

Three months after burying this black-headed gull's egg (arrowed) a fox has returned to its cache, dug up the egg and eaten it. Only the shell remains. The dug-up cache can be clearly seen (top-left), as well as fox prints (bottom-right and centre). The clump of sea-rocket plant was probably used by the fox as a landmark, but the exact location of the egg was almost certainly determined by scent.

He found this behaviour puzzling. What was the point of it? Did the foxes ever return to dig up their hidden spoils? Were these buried eggs really a store of food to be recovered and eaten at some later date? It seemed the obvious answer. And yet during all his tracking sessions day after day during the breeding season, he had found no evidence of this. At no time did fox prints show that the animal had returned to the scene of its plunder to reclaim its 'loot'. It was not until he returned at the end of July to spend a holiday in camp among the dunes that the reason for this became clear—he had been looking at the wrong time of the year!

Now, in late summer, the migrant seabird population had long since flown away and the dunes were almost deserted. But fox prints leading to scuffled sand with pieces of egg-shell round about, proved that although their main source of food had gone the foxes were coming into the dunes at night and digging up some of the eggs they had buried three months earlier. Clearly, these eggs buried during the nesting season *did* provide a supply of food, at least for a time, during the lean days after the black-headed gulls and other ground-nesting birds had left the peninsula.

But this discovery raised a very interesting question: *How did the foxes find the eggs?*

Tracking showed that no digging had been done at random; a fox dug only where eggs had been buried. This seemed to indicate an astonishing acumen. Since April or May, when the eggs had been buried, sand storms had long ago covered all traces of the original fox scrapes. In such a vast area of sand and gravel and marram, how did the foxes know so precisely where to dig? What sense led them unerringly to the exact spot in this waste of dunes where an object so tiny as an egg was hidden?

Nobody knew the answer, although the general opinion was—memory.

Tinbergen thought otherwise. He conceded that memory probably guided a fox to its previous hunting area, but considered it unlikely that memory alone was responsible for pinpointing an egg so exactly. He suspected that what led the foxes to the eggs was a sense of smell—even though after many sand storms the eggs might be underneath several inches of sand.

In order to solve the problem, Tinbergen decided to carry out a simple experiment, aided by Dr Monica Impekoven from Basle University. They argued that if a hungry fox really could scent the eggs it had buried itself, it should also be able to scent eggs that *they* had buried—and which, obviously, the fox could not possibly remember.

And so, during one nesting season, upwards of a hundred hen's

Preparing the egg-line. Dr Monica Impekoven with Niko Tinbergen (left) and the author.

eggs were buried in a long line across the dunes. This was done in an area regularly visited by foxes and where they tended to bury some of their stolen eggs.

Intrigued by this experiment, and realizing the possibility of filming a most original and dramatic ending to *The Sign Readers*, a wildlife picture about nature detective work that I was making for the B.B.C. Natural History Unit, I decided to stake everything on the fox finding the eggs and to use it as the highlight of the film.

The eggs were buried during May, and every day from then on the egg-line was inspected for signs of foxes.

It had, of course, been considered possible that disturbance of the sand caused when a fox buried an egg might, later, have drawn attention to the spot. This seemed unlikely, owing to the amount of drifting sand. Nevertheless, an equal number of empty scrapes had been made. These alternated with the scrapes containing eggs, each scrape being about three yards apart. *Now* if a fox dug only where eggs were buried and left the empty scrapes untouched, soil disturbance was not likely to be responsible.

Several yards from each scrape a numbered marker peg was put in the sand, with a stone beside it. This was of the greatest importance. Wind-blown sand smoothed out all the scrapes shortly after the eggs were buried; only the pointing markers showed their positions. By now, the eggs were several inches deep.

All this having been duly filmed, I sat back and waited for something to happen. If buried eggs were found and dug up, we had a highly successful ending to the picture. If not, I was faced with a disastrous anti-climax. Everything depended on the fox!

During May and June, while there was an abundance of food for them in the gull colony, the foxes showed no interest in either the hand-buried eggs or any that they had buried themselves. This we had expected, and although time was drawing on I was not unduly alarmed.

But by mid-July, although the gulls had left the dunes and flown out to the beaches with their young, and hunting had become much more difficult for the foxes, the egg-line was still undisturbed.

I had hoped to have the film cut and dubbed before the end of August, and day after day we eagerly inspected the egg-line ready to film the longed-for result. But although tracking proved that foxes were continuing to use the dunes, the eggs remained intact.

And then in the early morning of the last day of July came a moment of great excitement—during the previous night the first of the hidden eggs had been dug up and eaten.

But not by a fox.

As tracking quickly proved, this nocturnal egg-finder had been a hedgehog!

That the hedgehog had found the egg by scent was equally certain. Its line of tracks leading across the sands and then curling round and round the scrape, showed how the animal had first winded the egg, then wandered round it in circles until directly overhead. Then, its exact position pinpointed, the egg had been dug up.

This incident, besides being of great interest in itself, raised my spirits. If a hedgehog could find a buried egg by smell, a hungry fox should certainly be able to do so. It was surely only a matter of time.

And so it proved. On the second night of August a fox returned to the deserted sands and dug up two of our eggs. There was little to see, the signs were slight enough; a scattering of sand, a few fox prints, some fragments of egg-shell; but the story they told was sufficiently eloquent. A rich reward for the many weeks of patient tracking.

From then on the fox's visit became a fairly regular occurrence. Empty scrapes were ignored. In every case the only places disturbed were those in which eggs had been hand-buried.

The issue was beyond doubt. The fox could have found those eggs only by sense of smell. It had solved the riddle with its nose and added a fascinating piece of information about its hunting behaviour.

So, too, had the hedgehog!

There was an interesting sequel to this. Later that year, Jeffrey Boswall of the B.B.C. Natural History Unit, producer of the famous *Look* series, was staying with me at Cragg Cottage. The buried-egg experiment had created much interest, and one grey December afternoon we set out across the peninsula towards the dunes where I had filmed during the previous summer. Soon, we found ourselves following a trail of fox prints that led towards the sandy valley of the egg-line.

Winter gales had blown up ridges of sand, so that few of the marker pegs remained visible. Some, however, could still be seen and from these it was possible to reconstruct visually the line of handmade scrapes.

Our fox trail crossed and re-crossed this line—until, suddenly, a fresh disturbance in the sand together with tell-tale pieces of shell showed where an egg had recently been dug up and eaten. So—a few untouched scrapes remained. And one fox at least was scenting and unearthing eggs that had been buried nearly seven months before.

From start to finish the fox itself had never been seen. The story of its activities leading to new insight into its behaviour had been constructed entirely by nature detective work.

Kill by night—and day . . .

The hedgehog moves into the dunes during the second half of April when the eggs are being laid. It has a varied diet: beetles, slugs, worms, mice and frogs, among other prey; but, contrary to much that has been written about this animal, it is also a menace to the smaller species of ground-nesting birds. In a black-headed gull colony on dry ground it is a formidable predator. After roaming the dunes at night in search of food, it hides during the day in thick cover, or in holes, and the story of its hunting activities must be pieced together by sign reading. On grassland this is not so easy, but in the dunes where, on a still morning every footprint or scuffle remains clearly printed in the sand, the truth is plain to see.

Hedgehog tracks on dune slope close to the estuary.

Hedgehog prints in sandy mud. At night, hedgehogs sometimes travel for long distances in search of food, and at a surprising pace. But they have only one gait for all speeds.

Opposite
The purposeful 'trod' of a hedgehog that would occasionally visit a research camp on the sand dunes during the night and raid the larder! As one camper remarked: 'You could hear it rattling about. It would take scraps of food left on plates and in saucepans and in a refuse pit.' An interesting sidelight on hedgehog behaviour. We can see where this animal has set off again on its long journey back across the estuary sands at low tide.

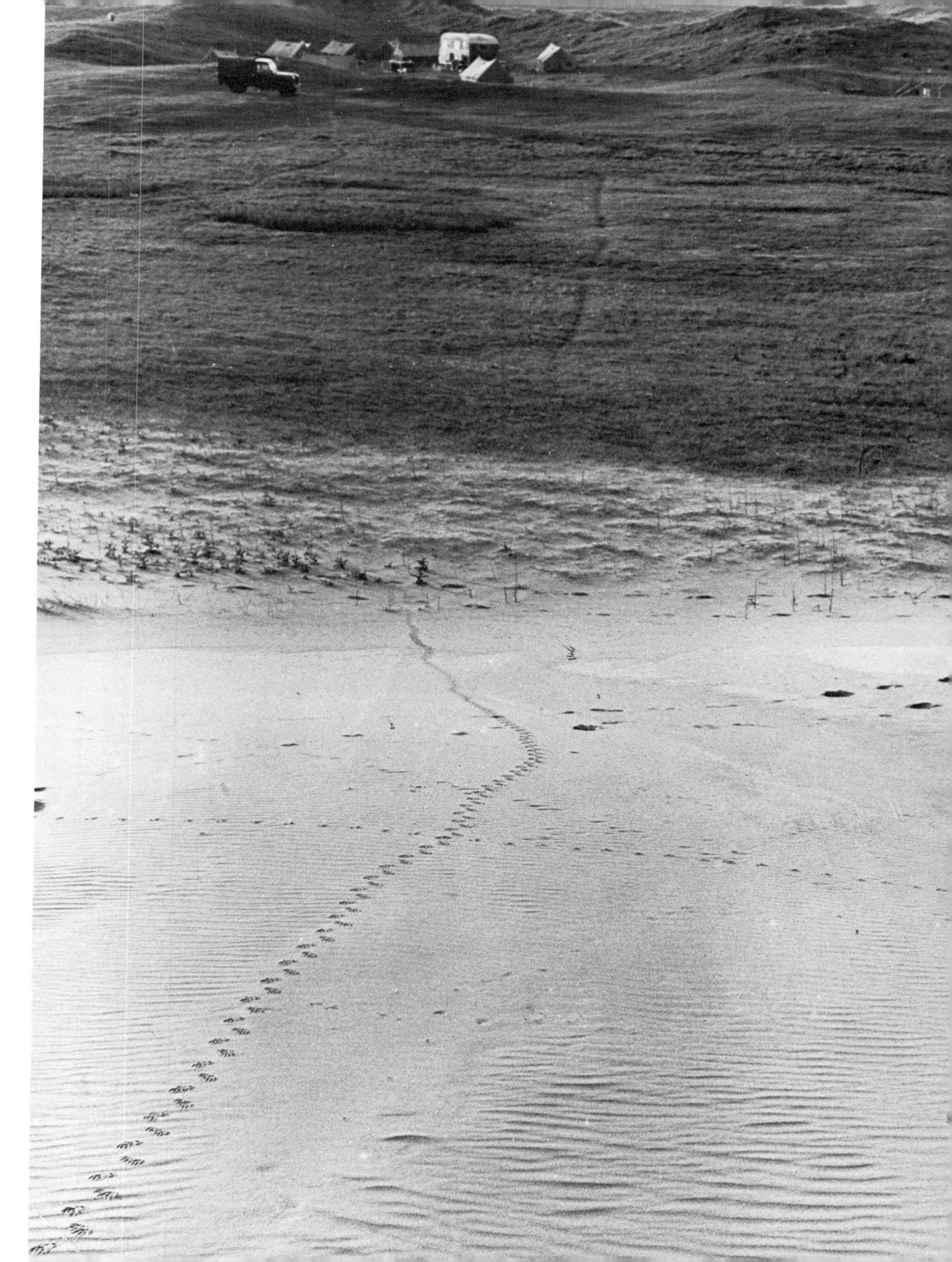

Daylight attack
A herring gull has killed a black-headed gull chick, but has, it seems, been driven off before it could eat its meal.

A hedgehog will eat up to ten eggs a night, and it is commonplace to find a line of hedgehog tracks leading to the nest of a gull, tern or ringed plover, with bits of egg-shell scattered further on along the trail.

But as well as stealing eggs it will attack the chicks—holding them down and munching them alive, starting at the rear end. The screaming of young birds being slowly gnawed to death can sometimes be heard at night. Subsequent investigation will disclose the unmistakable evidence of hedgehog predation: dead or badly mauled chicks, bitten about the legs and anus.

As will be known by anyone who has walked through a colony, the gulls have a habit of defecating on intruders into their territories, and the hedgehog is no exception. Hedgehogs are often found with tell-tale white blobs on their backs—sure signs that those animals are no strangers in the gullery.

The gulls' response to the hedgehog is different from their reaction to fox or stoat, when a number of birds will combine to defend a large area. Against the hedgehog, each bird seems content to defend only the small area surrounding its own nest. Other gulls, attracted by the nesting bird's alarm call, come along to see what all the fuss is about. They attempt no mass attack, but settle a short distance away to watch events. Reluctant to expose her eggs the nesting gull remains crouching on her nest with wings outstretched. She seems tempted to fly at the intruder and attack with her beak, but will only do so if the hedgehog ventures too close inside the territory. At least, this has been the gull's behaviour on those occasions when I have witnessed such an event.

When hard pressed, the hedgehog's defence reaction is to protect its head by curling up into a spiky ball. This, presumably, is why so many are killed by cars on the roads at night.

Strangely, although herring gulls are less sensitive to attack than carrion crows, only a few invade the colony—even though a large number of these big gulls quarter up and down the nearby sands. But eggs or chicks in outlying nests stand no chance of survival. Isolated pairs of black-headed gulls cannot defend their nests against either herring gulls or crows.

Crows are easily driven away from nests in the colony itself by the gulls' social attack. The crow is a hated enemy. If one appears over the main nesting area, the gulls will rise and attack it in force.

Occasional gulls seem to be 'crow mad'. Once, when returning to camp after filming with Hans Kruuk—who was studying gull attacks on dummy herring gulls and crows—we left a stuffed crow standing on a nearby grassy hillock. While in the middle of our coffee break in the camp caravan, an outcry from an angry black-headed gull brought us to the window. A solitary gull was making noisy and repeated attacks on the dummy, swooping low and striking with its feet. Feathers flew from the crow's head and back as it rocked on its heavy stand.

Quickly setting up a camera we succeeded in filming a number of attacks—which ended only when the screaming gull, in an even more frenzied assault, overturned the crow, stand and all.

The bird made no further attacks. Honour seemingly satisfied, it shook itself, then headed off in the direction of the distant gullery.

A riddle of the sands

An empty egg shell lies by itself on a bare, unmarked stretch of open sand. There is no problem of identification: it is the shell of a black-headed gull's egg and, as shown by the membrane and streaks of blood inside the shell, a chick has recently hatched from it. If a crow or some other predator had rifled the egg, the shell would have been all white inside.

So—we have a gull's egg that has hatched a chick. But how did the shell arrive on this patch of unmarked sand? There has been no wind to blow it here. A chick cannot have hatched in this spot: there are no tracks or marks of any sort round about—although the sand is soft enough to hold them—and there has been no rain to obliterate any tracks that might have been left. So, how did the egg-shell get here?

A nice problem for the nature detective.

'You know my methods,' said Sherlock Holmes. 'Apply them . . . How often have I said to you that when you have eliminated the impossible, whatever remains, *however improbable*, must be the truth?'

And in this case the truth is not far to seek. Since the shell cannot possibly have arrived overland or underground, it must have come by air.

And it did.

When the chick had hatched, the parent gull took the empty egg-shell in its beak, flew off with it and dropped it here, well away from the nest site.

Many species of birds do this: the oystercatcher, for example. It is behaviour that has an important function. Unbroken, gulls' eggs are beautifully camouflaged and almost invisible from above. But broken shells show up very plainly. The flash of an empty egg-shell is similar in effect to the flash of a white face. Soldiers on the ground are difficult to spot from the air—until they look up.

If empty shells were left in the nest after the chicks had hatched, they might quickly attract the attention of a wandering herring gull or carrion crow.

Egg pecked open by crow. But the thief has been disturbed before the meal could be finished.

Strange adventure

Sand dunes and the sea shore seem unlikely places to find badgers, and yet their tracks have been found there on a number of occasions. Being almost entirely nocturnal they are seldom seen except at dusk and daybreak, or in a car's headlights. Much of what we know about their hunting behaviour has been built up by nature detective work. They are very much creatures of habit and will cover long distances at night, tending to use regular 'trods'—a characteristic well known to the forester, who saves his rabbit fencing by building special swing-doors to allow a badger to get in or out of the forest without forcing its way through the fence.

Badgers have an astonishingly wide range of food. Much of their diet consists of grubs and roots, but in addition to insect and vegetable food they certainly have a taste for eggs and flesh. They have been known to feed on eggs and birds, both adults and young, in the gull colony, and I have found their tracks during February and March, when they have been hunting in the dunes for rabbits.

Opposite
The characteristic footprint of a badger (arrowed). Very broad and bear-like, it cannot be mistaken for the print of any other European mammal.

A rabbit 'stop' dug out by a badger. The young rabbits have been decapitated, their bodies eaten.

Close-up of the badger kill. Nipped-off heads are typical of badger kills. This particular badger, tracked by Niko Tinbergen and myself during the making of our wildlife picture *Tracking*, had dug out and eaten the inhabitants of two rabbit stops, in addition to raiding several black-headed gulls' nests, during the course of a single night's hunting.

Opposite bottom
Badger tracks on the seashore. The 'bulldozed' heap of sand shows where the animal has been rooting with its snout—which it uses to grub-up shallow-lying bulbs and insects. For deep-lying food it uses its powerful paws (see dug-up rabbit 'stop', p. 81).

One night at the height of the breeding season some years ago, a badger—seemingly from the wooded slopes several miles inland—made its way down to the shore and swam across the shallow estuary at low tide. As its line of footprints revealed, it headed straight across the broad sweep of sands and finished up in the gull colony (in foreground of picture).

Further tracking showed that the badger stayed in the dunes for several weeks. It fed mainly on eggs, gulls and their chicks, and young rabbits, and slept in an old rabbit hole that it dug out and furnished as a temporary burrow. On several occasions, observers in a hide watched it preparing its bedding at dusk.

Above
Badger prints were found at the spot where the animal had landed on the near side of the channel (arrowed).

The badger digs a hole in which to defecate. This lavatory has been dug on the edge of the dunes, just above the tideline.

When the breeding season was nearly over the badger left the dunes—presumably for its original haunts—as mysteriously as it had come. An unusual example of badger hunting behaviour, discovered by tracking.

But how did the swimming badger know what lay ahead of it on the far side of the estuary? Why did it choose that particular night to roam so far? Why did it come only on that occasion? Badgers have been known to walk into the dunes from the landward end of the peninsula, but there is no record of any similar Leander-like adventurers.

Climax

The peak of the hatching period in the black-headed gull colony is between 15th and 31st May. It brings out a very unattractive characteristic of the black-headed gull—cannibalism. Like the herring gull and the lesser black-back, two other colonial nesters, the black-headed gull has a taste for newly-hatched chicks. The great respect in which the individual territories are held by the gulls inhibit wholesale killing of each other's young. But an unguarded nest site can prove too tempting. Every parent must be watchful of its neighbours. A wandering chick will be instantly snapped up.

It is interesting to observe a chick at feeding time. To get a meal the chick pecks at the tip of its parents's bill. This action stimulates the parent to regurgitate food, which it presents to the chick. The bill of the adult black-headed gull is red, and a chick instinctively responds to this colour. Experiments have shown that chicks hatched in an incubator prefer to peck at red objects when hungry. Since these chicks have never seen a living gull their begging behaviour must be innate. This is exactly similar to the begging behaviour of the herring gull or lesser black-back chick. In this case, the chick pecks at the red spot on its parent's lower mandible.

When they are three or four days old the chicks make what seem to be clumsy flying movements, jumping into the air and tumbling over on to their backs. It is very comical to watch.

They grow very fast and are soon able to take better care of themselves. If they stray into a neighbour's territory they will be harassed and viciously pecked—in which case they will make no attempt to stand their ground, but will run away home as fast as they can. But it is significant to watch what happens when the same adult neighbour comes into *their* territory. Although perhaps only a fortnight old, they will most resolutely hold their ground. Indeed, often enough, they will attack the intruder and drive it out. Respect for territory is deeply ingrained in every member of the colony. The intruding adult is in effect trespassing on 'forbidden' ground—and knows it. How very human it all seems.

Soon, the young gulls are trying to hunt food for themselves. They wander about pecking indiscriminately at anything they come across: rabbit droppings, stones, feathers, straws, pieces of stick, finding out by experiment what is edible and what is not. They even have a go at the obnoxious cinnabar moth caterpillar that lives on the ragwort.

Both in the moth and caterpillar stage, the cinnabar is most distasteful to insect-eating animals. The moth is coloured a glossy black with vermilion spots and bars. The caterpillar is a brilliant orange with black bars and provides an interesting example of 'protective colouration' as a defence against predators. The young gulls burn their mouths on the caterpillars' stinging hairs, and thereafter recognizing the cinnabar by sight seldom try it a second time. So—although a large number of cinnabars get killed, sufficient survive to maintain the species (see p. 29).

On warm June days the anomala beetle, a relation of the cockchafer, hatches in thousands. The gulls know this at once, for suddenly they are hawking beetles all over the dunes—a sign easily read from a distance by the nature detective.

At the height of the season the peninsula is teeming with wildlife and the valleys between the dunes are bright with flowers. Even among the waste of sand and marram there is a profusion of sedges and lichens with patches of ragwort, wild pansy, sea rocket

and silverweed. The spawning pools of the natterjacks are black with tadpoles. Young wheatears are peeping out of rabbit burrows. The air is full of the skylark's song, and snipe are drumming above the marshes. Waddling towards the estuary mud are the shelducks with their strings of ducklings—with the watchful herring gulls waiting their chance to snap up any youngster that lags behind. The terns are busy fishing, hovering hawklike fifteen or twenty feet above the water, before plunging vertically to snatch sandeels for their hungry young. Shoals of grey mullet that come into the estuary with every tide are swirling in the shallows. Returning salmon and sea trout leap as they taste the freshwater from their river of return. Cormorants dive for young sea trout and flounders.

Opposite
A study in concealment. There is little here that would catch the attention of the keenest-eyed predator. Even in close-up one has to look carefully to find the eye, the line of the beak . . .

Until, suddenly, the grass stems part—and the sitting snipe emerges from her nest in a tussock.

Cormorant prints.
Markedly inturned, even more so than those of the shelduck, cormorant prints show each of the four toes to be connected by the web.

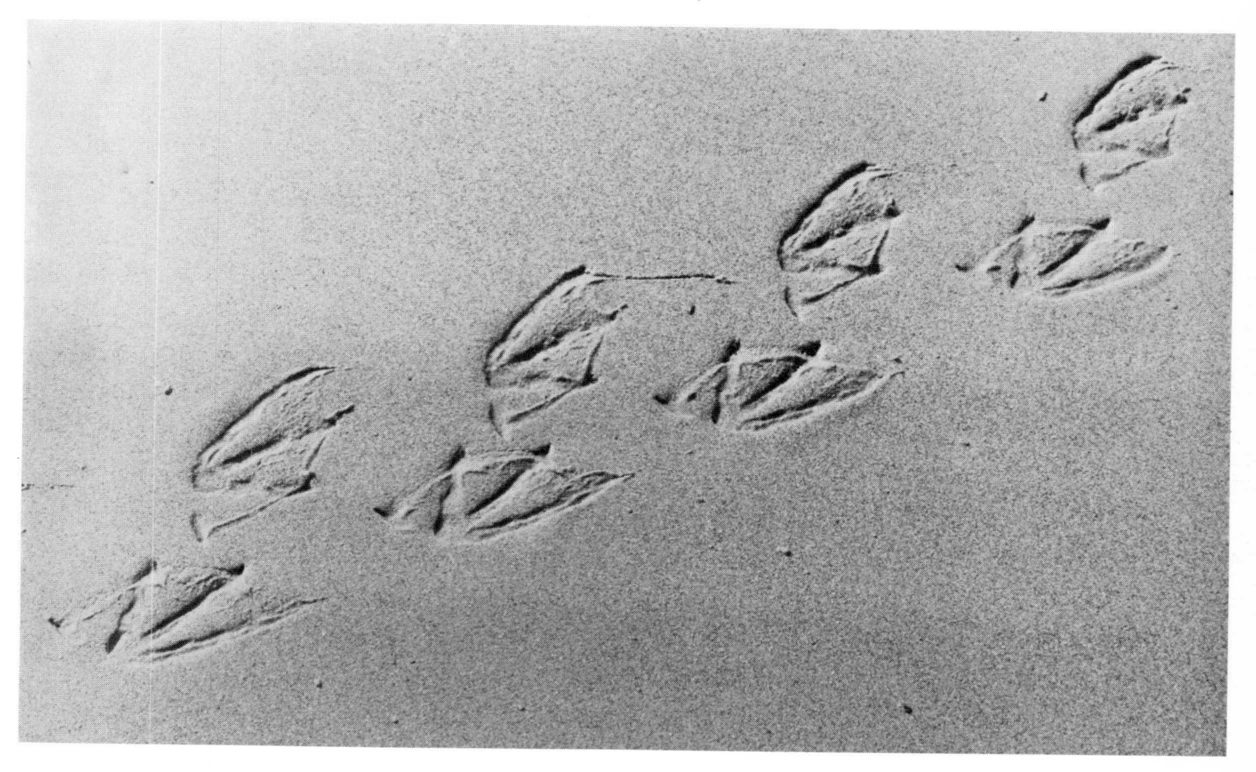

Family on the move!
Two parent shelducks (on the two flanks) have shepherded their brood of ducklings, from the nest site in a rabbit burrow, across the open sands towards the estuary.

Sheldock 'flapping' distraction display. The marks of the wings can be plainly seen.

When danger threatens, the parent birds feign injury, often running along with one wing trailing on the ground.

Desolate beauty of the dunes at the approach of winter.

As always, over all hangs the shadow of predation. In response to their parents' alarm calls, the young black-headed gulls crouch motionless under cover of the dense vegetation that has grown high around the nest sites. Even so, many fall victim to the fox.

But despite heavy losses, sufficient young birds survive to ensure the breeding stock of future years.

By mid-July the season is nearly over. The young gulls leave the colony and fly out to the distant shore. They are still being fed by their parents, and will return to their respective territories if the parents fail to come out to the beach to feed them. These young birds are still rather poor fliers, still very much in the 'learner' stage, often making crash landings downwind in a flurry of wings and legs. But they become more like the adults every day,

Herring gulls flying out to feed along a lonely shore at sunrise.

moulting their juvenile plumage so that the beaches become covered with chocolate-brown feathers.

And then, quite suddenly it seems, the gulls have gone. They leave together, and so strong is their social attachment that those adults that still have young in the colony fly off and abandon them.

There is a strange stillness. Shelducks and mergansers have long since gone to the estuary with their young. Oystercatchers and ringed plovers are out on the beaches. Soon the adult shelducks will be travelling across country on their moult migration to the tidal flats of the Heligoland Bight.

In August sunshine the dune vegetation gleams with patches of brown and gold. Insect-hunting natterjack toads are active at night, and long-tailed fieldmice enjoy the plentiful supply of marram grass seeds. Flocks of migrating golden plover drop in to rest on estuary sand banks. Southward flying Arctic skuas pause awhile to harass and rob the feeding terns as they plunge-dive off the shore. Soon the first migrant wigeon will come whistling from the north; perhaps a skylark-swooping merlin will flicker between the dune tops. But at night the empty nesting grounds lie silent under the stars.

We are back where we began. An autumn wind whines in the marram and once again the dunes are left to the sheep and the rabbits and the lonely fox.

Estuary and seashore

The huge variety of life that exists in estuaries or on the sea shore, or is brought in by a flooding tide and left stranded by the ebb, provides the food directly or indirectly for many species of animals—especially birds—a number of which live at least a part of their lives inland or among the sand dunes.

It is when they are 'on the feed' that most animals advertise their presence. At some time or other, albeit with varying frequency, all animals have to eat.

In order to bait his lines successfully, the fisherman must know what food is most likely to attract the fish he wants to catch. And the nature detective who wishes to study animal behaviour needs to know as much as possible about the food that his widely differing animals need in order to live and breed successfully. This knowledge comes partly from observation of the animals themselves, and partly from an ability to read the stories of their feeding activities.

Whether it is the sight of sandeel-hunting terns, plunge-diving over estuary shallows; a cloud of insect-hawking black-headed gulls wheeling high in the evening sky; the flight pattern of a shellfish-dropping herring gull, or simple signs such as the trail left in soft sand by the dragging body of a rabbit killed and carried off by a stoat, feeding animals seldom fail to provide us with evidence of their behaviour: evidence that, as we have already seen, can be read and understood by the astute nature detective.

Some of the animals that feed on the seashore have evolved amazingly skilled feeding techniques. To find out what the sea provides for these animals, and how they obtain it, is a study of absorbing interest.

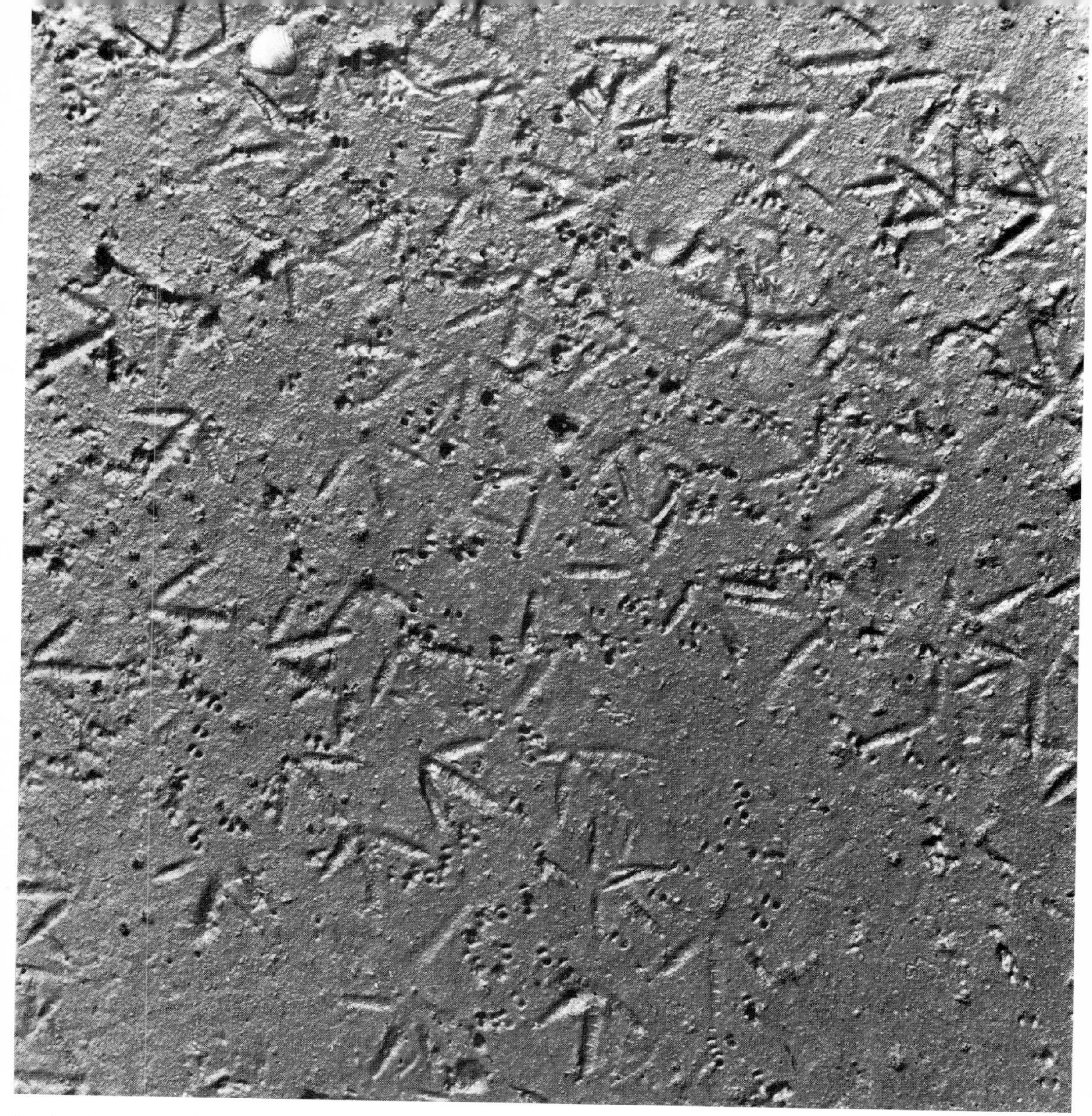

As the tide ebbs, a host of daylight feeders fly out to the seashore and the estuary banks, each leaving its own story in the sand or mud. Prominent among them are the waders, which hunt for food with specially adapted bills. The turnstone has a very short bill, which is used for flicking over pebbles or picking barnacles off the rocks. The sanderling, too, has a short bill and feeds mainly on or very near to the surface. The much larger redshank has a fairly long bill and probes for tiny shrimps and burrowing worms. The smaller dunlin—whose footprints and probing holes are shown in this picture—has a shorter bill than the redshank, but it is long enough to locate and pull out small ragworms. As can be seen from the 'double' probing holes shown, the bill was slightly open when pushed into the sand.

The tideline hunters and the hunted

Who can resist the lure of beachcombing? The profusion of flotsam and jetsam washed up on the tideline; the variety of creatures that hunt their living along the shore—and those that live there—provide a source of never-ending interest and fascination. Day by day, year after year, a number of animal beachcombers roam the stretch of shore between high tide and low tide. A study of their habits reveals how beautifully adapted these animals are to their environment.

On a stormy day with an onshore wind sweeping the breakers in towards the coast, we glimpse just a tiny fraction of what the sea contains. Whether floating, or brought in from the sea bed, there is always *something* being washed ashore. For a child—and many a parent as well—there is the excitement of fishing out a piece of wreckage, an old lobster pot or fish box; finding the first razor-clam or cuttlefish bone or bunch of bladderwrack or, perhaps, that strange, black, horned object known romantically as a 'Mermaid's Purse', but which is really the egg-case of a skate.

Whatever our age, a walk along the beach is an ever-fresh delight—with wonderful opportunities for the nature detective. In addition to an extraordinary variety of man-made objects, the waves wash in a huge quantity of marine life, both plant and animal, and spread it out dead or dying along the tideline.

At first sight this over-spill of life seems to be a complete loss—as though, by some strange oversight, nature has blundered. But we soon find that these accidental spoils of the sea are by no means wasted. Indeed quite the reverse is true. Far from being wasted, this mass of dead and dying marine life is put to very good use by an army of animal beachcombers who, in the long course of their evolution, have adapted themselves to living off the shore largely on what the sea provides. For them, the littoral—the narrow strip of shore between high and low water marks—is far more than a casual boundary between dry land and sea.

Nothing goes begging on the sea shore. Many food chains, based on what the sea discards, merge to form a living pattern in which even the stranded seaweed plays its part—for it harbours hordes of sandhoppers. Lift a clump of weed along the high water mark and watch them leaping like fleas, before burrowing quickly into the sand again.

Sandhoppers live mainly on decaying seaweed, but they also clean the bones of dead sea creatures washed up on the tideline: perhaps the skull of a dolphin; a gannet's carcase, or the skeleton of a baby seal. And the scavenging sandhoppers are preyed on by a host of waders: the whimbrel, for instance, that feeds mainly along the high water line, together with turnstones, ringed plover and many others.

Successive high-tide lines, showing spring tides dropping towards neaps.

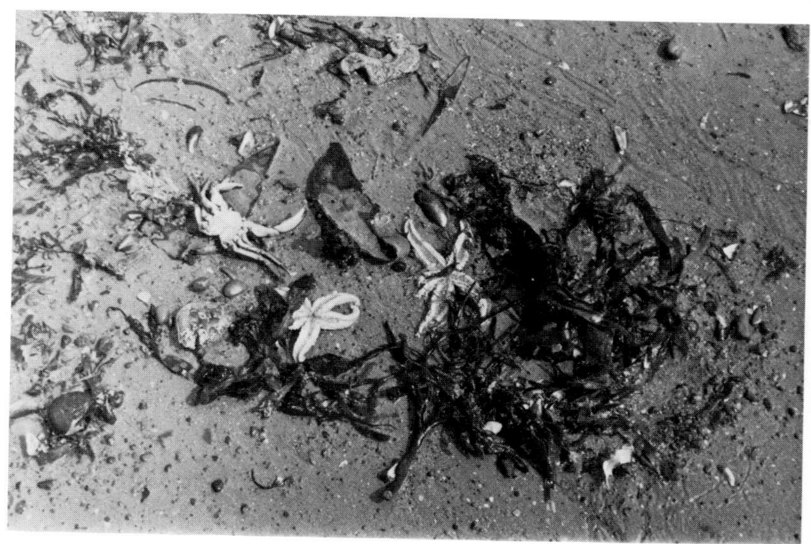

'Over-spill of life' along the tide's edge.

The starfish lives almost exclusively on mussels and clams—which it opens by suction from its five 'arms'. The digestive juices in the starfish's stomach do the rest.

At low tide the beach seems empty. But hidden away beneath the sand and mud are crustaceans, shellfish, sandeels, worms; their presence betrayed only by the signs they leave behind them on the surface. The ebb and flow of the sea is of immense importance, both to these creatures and the creatures that prey on them, because the rising tide washes in another form of marine 'waste': plankton—the almost invisible food of the sea.

Plankton—microscopic plant and animal life—provides nourishment for the sand and mud dwellers. On it, all marine animals depend for their existence.

Each flood brings in a vast amount of plankton, which is filtered out by millions of shellfish, worms and other animals. Some creatures, such as the marine worms, filter this plankton through the sand. Bivalves, such as cockles or mussels, or razor clams, filter it directly from the seawater. And these animals, in turn, provide food for larger animals.

Many different beachcombing species can live together on the shore without competing with each other, because each has

Sea urchin's skeleton.

Egg-case of a skate.

developed its own feeding techniques and skills. And the beaks of the various beachcombing birds have evolved to suit these different requirements.

Some birds find their food on the surface: the gulls, for instance, and that beautiful little wader the ringed plover, that can so often be seen trotting along delicately picking up tit-bits beside the tide's edge.

Other waders, such as the knot, have slightly longer bills and bore down a little way, about an inch or so, to find food that is just

Masked crab.

Opposite
Like some weird plastic jellyfish, a 'seabed drifter' lies stranded among a tangle of bladderwrack. 'Drifters', which bump along the sea bed until finally cast ashore, are used by marine scientists in their study of ocean currents. Each carries a number, and an address to which the drifter should be sent. The finder is rewarded.

Above left
A cluster of shells collected at random from the tideline from one spot within arm's reach.

Above
Whelk egg-cases.
The whelk is a carnivore that eats food left by other predators.

below the surface. The knot—or, to use the Continental pronounciation: '*k*noot'— is supposedly named after King Canute who, so history relates, also visited the tideline. *Not*, as has so often been said, in an attempt to stem the rising water, but to prove to his sycophants the absurdity of his trying to do so.

Longer billed waders, such as the oystercatcher and godwit, probe for food that is lying deeper still. The curlew, largest of our waders, probes deepest of all. Female curlews (whose bills are longer than the males') can probe some five inches deep and catch worms inaccessible to other species.

It is not only the feeding gulls and waders that leave signs of their hunting activities. Their prey—the tiny shrimps and snails and worms—leave their own feeding tracks. Often enough these tracks betray the little animal's presence in a particular patch of sand or mud, and it is likely that the predators take advantage of this. Observation of wader feeding behaviour leads one to suppose that the birds look for outward signs of the prey's presence. They, too, it seems, are capable of doing some 'nature detective work'. Once they have probed below the surface their prey—which, of course, is out of sight—is detected by touch but if sign-reading has guided them to an area likely to hold food their probing is not quite so haphazard as it might seem.

At low tide, a walk on the beach near the tide's edge or across the sandy mud of an estuary, usually takes us past a multitude of lugworm casts. The lugworm makes a U-shaped burrow, open at the surface, and lies head down inside it. Sand and organic matter are ingested together, the sand being discharged in little coils (or 'casts') from the open end of the burrow.

Nourishing food for many seashore and estuarine animals. Hydrobia (approximately life-size) surrounded by their feeding tracks, just after the tide has left these tiny shellfish high and dry.

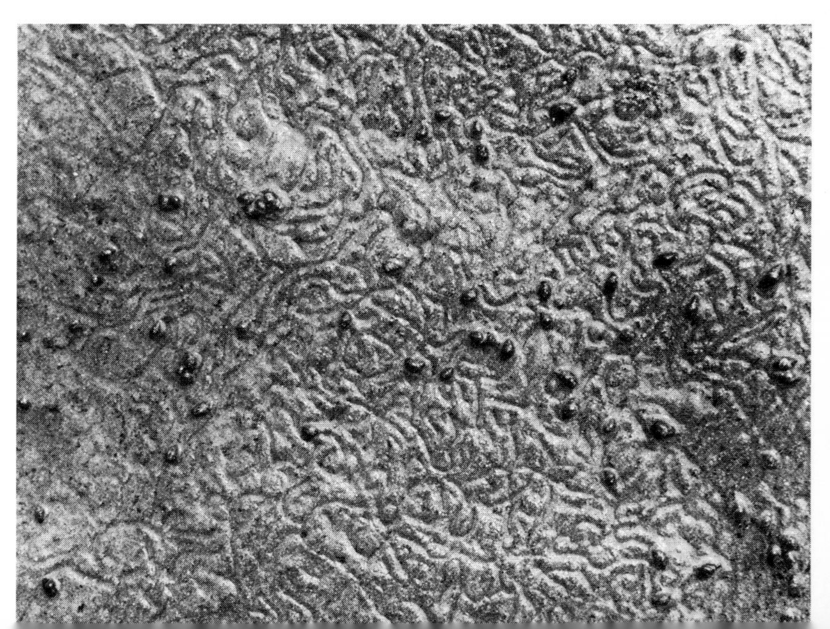

Left
Double probing-holes made by a snipe in estuary tide-line mud.

Below
The 'trampling marks of a black-headed gull in soft sand. This rather curious method of obtaining food is also used by the larger gulls: herring gull and lesser black-back. Trampling (paddling up and down with the feet) is often done in shallow water. The suction caused by the two webbed feet brings food up to the surface: amphipods, bristle worms, hydrobia—tiny winkle-like shellfish.

In this instance, dunlin prints surround the trampling site. Clearly, the little wader has taken advantage of food drawn up by the gull.

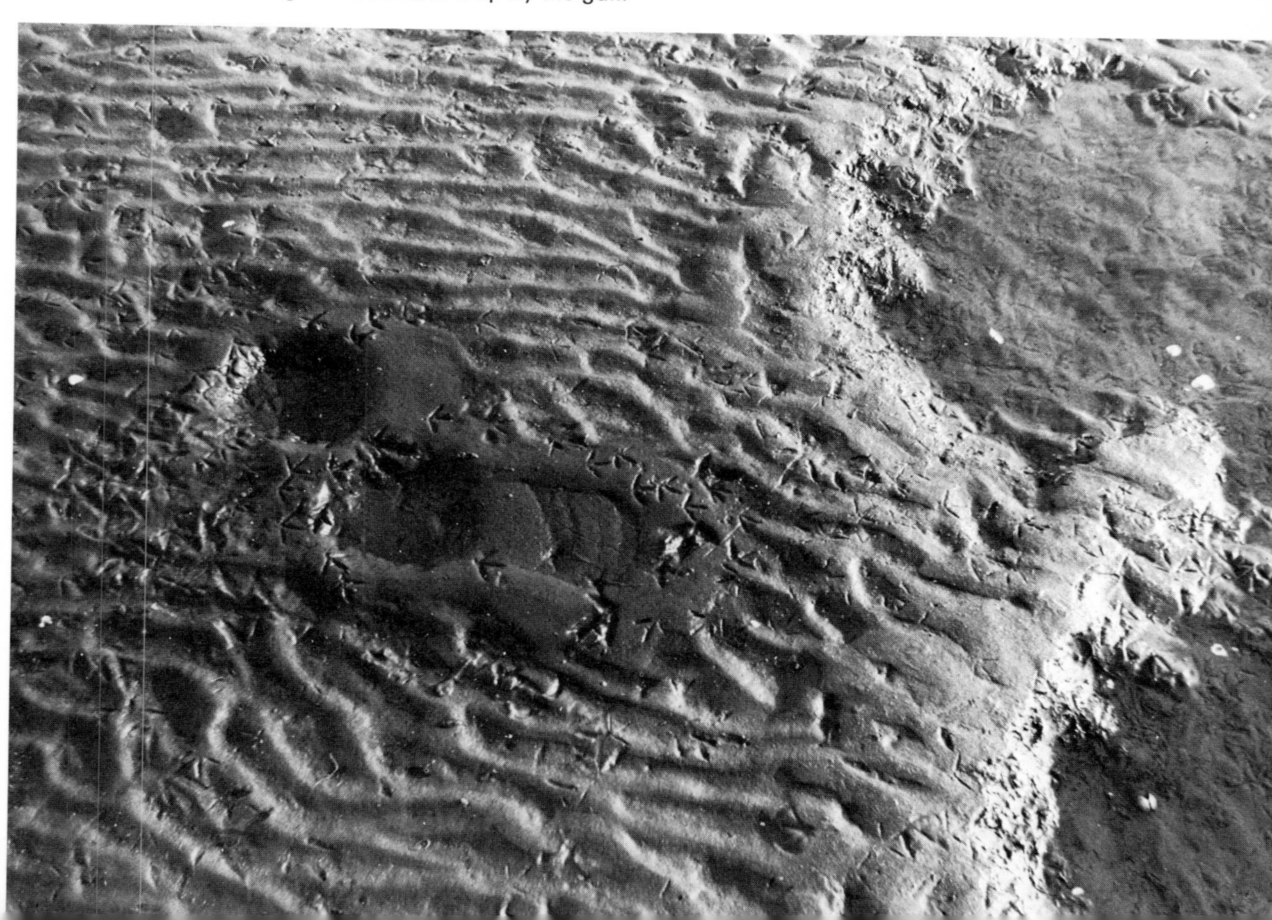

Close-up of a black-headed gull's trampling marks in mud, showing a preponderance of hydrobia that have been brought to the surface. Also shown: cast of a small lugworm (arrowed).

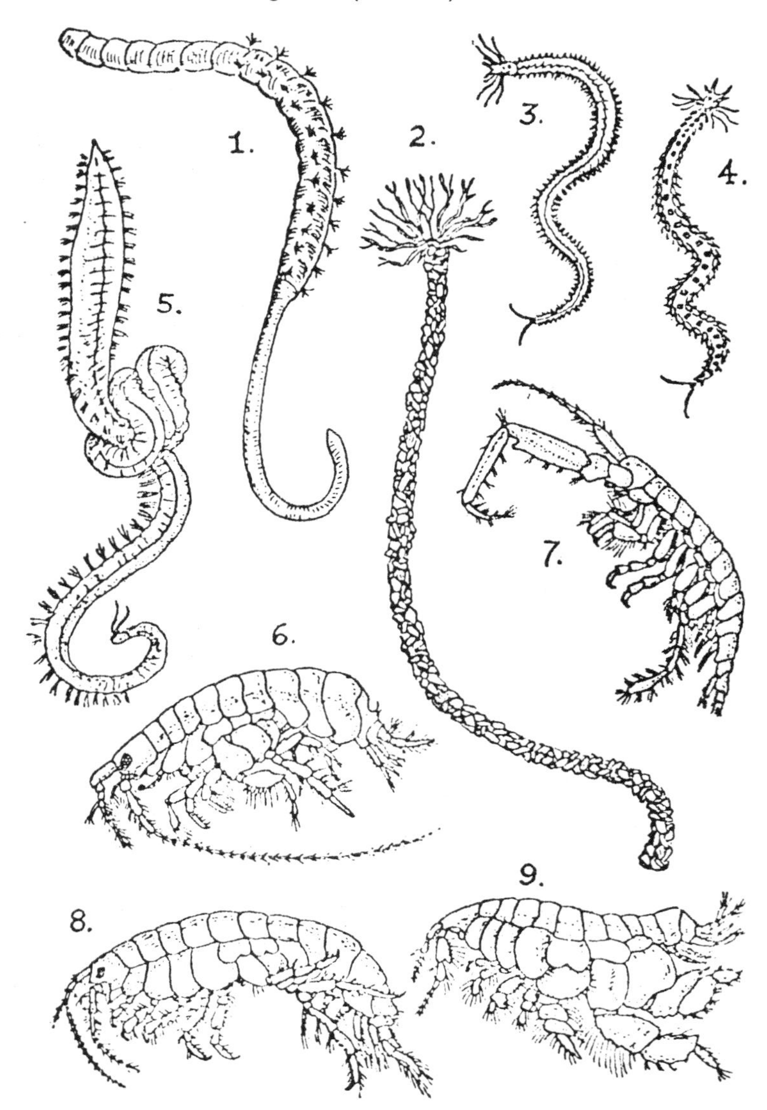

Some sea-bed fauna, including lugworm 1, sand-mason 2, ragworm 3 and sand-hopper 6.

After Barrett & Yonge's 'Pocket Guide to the Sea Shore'

Bar-tailed Godwit.

'Godwits must have been considered a good omen, if their name represents, as seems possible, the O.E. *gōd wiht*, "good fellow", a title given also to the kindly sprite, Robin Goodfellow. Attempts have been made, however, to connect the name with the edible qualities of the wader, which was once a favourite tablebird.'

Stephen Potter and Laurens Sargent, *Pedigree: Words from Nature*.

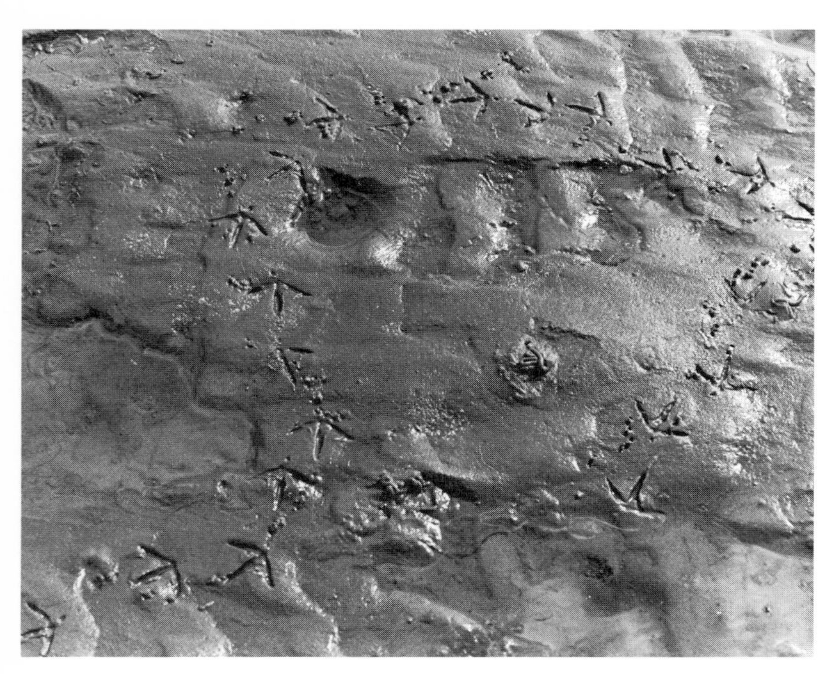

Top
The prints of a bar-tailed Godwit circling a lugworm cast and hole. Other lugworm casts dotted around.

Like that of many other polychaetes (worms having numerous bristles along their middle sections) that inhabit places deficient in oxygen, the blood of the lugworm is rich in haemoglobins. This enables the worm to extract sufficient oxygen for its needs (through the frilly gills that extend along its sides) when there is very little oxygen present in the water of the burrow.

Bottom
As the tide ebbs, water drains from the beach through innumerable sand gullies. Here the bank of a gully has collapsed, exposing two lugworm burrows (arrows 1 and 2); a cockle (arrow 3), and a peppery furrowshell (arrow 4).

Above the place where the mouth of the worm is working, a cone-shaped depression forms in the sand. By noting the position of cone and cast, the bait-digger can tell at a glance the direction in which the body of the worm is lying. This enables him to align his fork and dig up the worm without cutting it in half. An example of very elementary applied nature detective work.

As the ebbing tide runs out of the tidal creeks and river estuaries, leaving saltings and mudbanks uncovered, out come the gorgeously coloured shelducks to feed. They get their food by skimming off the surface film of mud and extracting tiny shrimps and snails, such as hydrobia and corophium. Their bills, slightly up-tilted at the end, are splendidly shaped for this method of feeding. The tracks they leave behind them while feeding are weird but unmistakable.

Shelduck prints in estuary mud, with the bird's strange scythe-like feeding marks, made by the bill as it sweeps from side to side sieving up food. Also to be seen are black-headed gull and dunlin prints, hydrobia, and the star-shaped feeding patterns of small ragworms.

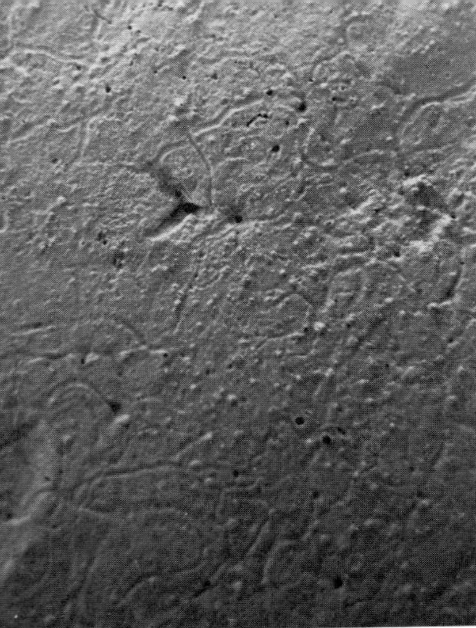

Left
Hole left by the feeding peppery furrowshell in soft mud.

Below
Feeding patterns left on the sand by tiny ragworms. The channels show where the worms have temporarily vacated their burrows and made sorties across the surface. Also shown is a shell of Baltic-tellin.

Opposite top
Spatter marks on the sand, made by a small clam (*Mya arenaria*) when it ejects water through its syphon hole (arrowed). At bottom left are three lugworm casts and the tracks of some small species of winkle.

Opposite bottom
Periwinkle tracks. Research has shown that the periwinkle can orientate itself by moving towards and from the sun. Hence its oval route—which tends to keep the animal in the same situation on the sea shore.

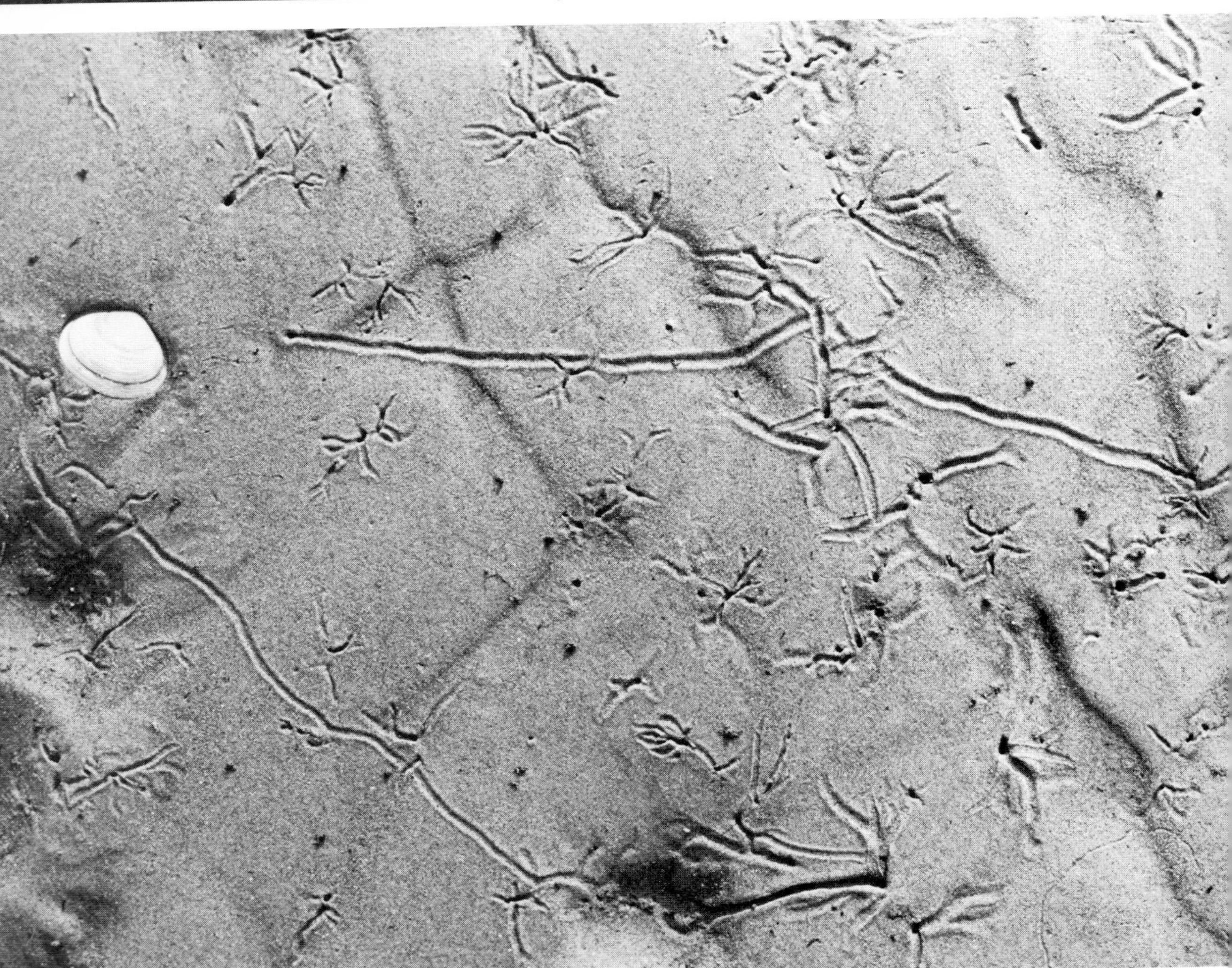

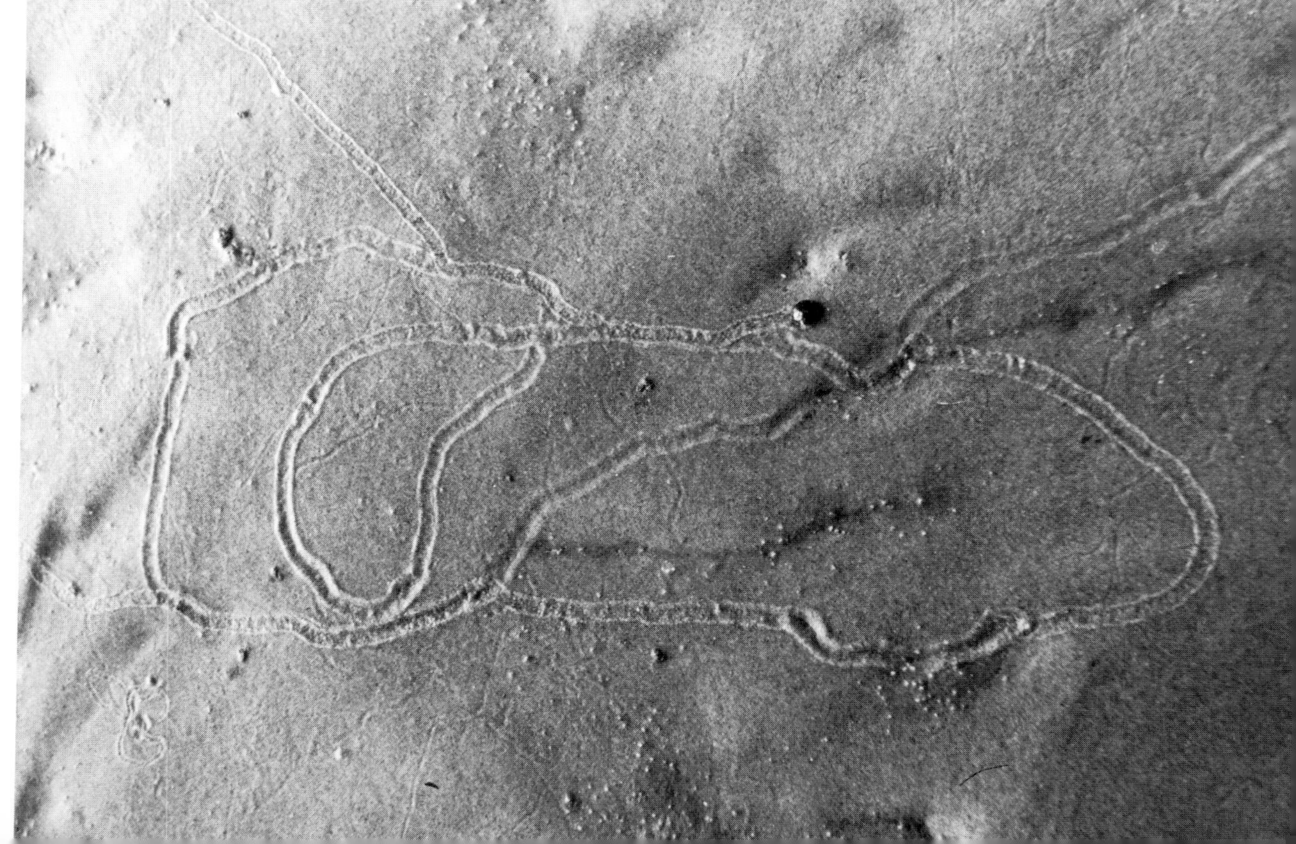

A friendly, rather comical character, the oystercatcher is a very common wader. Its twittering, piping call is heard almost everywhere along the coastline. The black and white plumage and red legs make it a very striking bird, but its title is something of a misnomer. It probes for worms and other creatures, and also uses its long orange bill most expertly to open various shellfish—but not oysters. Perhaps because in Britain it doesn't get the chance.

Features of the oystercatcher's bill that distinguish it from the bills of other waders are its size and blade-like shape. In addition to possessing sensory corpuscles near the tip—a basic characteristic in the bills of birds that probe for food—the oystercatcher's bill is specially adapted to withstand the stresses incurred when shellfish are prized open. Its triangular-shaped structure gives the bill great strength and robustness.

The specialist

At first sight a feeding oystercatcher seems to be taking 'pot luck', haphazardly pushing its bill into the ground here, there and everywhere in the hope of finding a hidden worm or shellfish. But is the bird really probing at random? Is the finding of a food item purely a matter of chance? There is evidence which suggests that detective work may also play a part in the bird's hunting behaviour and, indeed, be responsible for much of its success.

An example of this has been recorded by a scientist who spent several years studying oystercatchers. One afternoon, from his observation hide far out on the tidal flats, he noticed an oystercatcher following the line of footprints he himself had recently left in the mud while on the way to his hide. The bird was going from footprint to footprint, probing in every one of them and finding food in many.

It was, of course, very amusing to see that one of the birds he was following was, in turn, following him. But later, when he went to investigate what the bird had been up to he found that it had been feeding on peppery furrowshells. Often these shellfish lie too deep in the mud for an oystercatcher's bill to reach. But in this case the depth of the shellfish had been reduced by the depth of the footprints, so that the bird had been able to reach them and hook them out quite easily.

One astonishing aspect of this observation was the speed at which the bird had learned to probe in the footsteps. It seemed to show that oystercatchers can quickly take advantage of various signs. In this case the sign was not the mark left by the feeding syphon of its prey (see p. 109), since this would presumably have been obliterated by the footprint. What attracted the bird, it seemed, was the pronounced declivity—which promised an easier probe for anything lying underneath. The bird was probably probing 'blind'. But it seemed to realize that if there *was* a prey species in the mud beneath a print, it would be more readily accessible than it would be if probed for on the higher level of the surrounding flats.

In winter the oystercatcher is a social bird, gathering in huge flocks on the flats of river estuaries. But in spring and summer it is solitary and fiercely territorial. The nest is a slight scrape in sand or shingle, usually on the open shore above the high-tide line; but some birds nest well inland among the shingle of freshwater pools on the upper reaches of a river. Oystercatchers return year after year to the same nesting areas.

A pair of birds will make a number of nest scrapes, in one of which the eggs are laid. The eggs—there are usually three—are very well camouflaged and blend beautifully with their background. When the chicks hatch, this camouflage would be

Oystercatcher beak marks in sandy beach show where the bird has been probing for tiny worms and shellfish. It is interesting to see how the bird has made its task easier by choosing to probe mainly along the lower levels of sand.

destroyed by the white of the empty eggshells if they were left in the nest cup. But this doesn't happen. A parent bird immediately flies off with an empty shell and drops it well clear of the nest.

Like that of all ground-nesting birds, the oystercatcher's nest is vulnerable to predators such as the carrion crow. But crows by no means have things all their own way. A parent oystercatcher will defend its nest most resolutely against crows, rushing at an intruder with its long sharp bill stretched forward like a lance. It is

Top left
Holes of peppery furrowshells with syphon feeding marks. The peppery furrowshell's curious star-shapes are made in the mud by long snorkel-like feeding tubes through which, from a depth of six to ten inches, the buried animal 'hoovers' in the surface film and filters out food (plankton) left by the tide.

Top right
Tracks left by oystercatcher as it walked across estuary mud at Ravenglass in search of peppery furrowshells. But what animal has left the larger, rounded tracks that stretch from right to left? (For answer, see p. 115.)

Above
Peppery furrowshell dug out from a depth of 7–8 inches.

seldom that the crow does not beat a hasty retreat.

Young oystercatchers can walk almost as soon as they are born. Within a couple of hours of hatching, chicks born on the shore leave the nest site and accompany their parents out on to the tidal flats. There their parents feed them until they have learned to fend for themselves. The open flats would seem to offer little in the way of protection for young birds, but the camouflage of the oystercatcher's juvenile plumage provides comparative safety.

The sign of a 'Hammerer' (see text).

A prized-open mussel-shell—evidence of oystercatcher feeding behaviour. A portion of the shell has been broken away, and the adductor muscle, which held the two halves of the shell together, has been severed. The mussel flesh has disappeared.

The picture sequence (left and opposite) illustrates the feeding technique of a 'Stabber'.

Knife blade has been inserted between the two shells. It is now in the act of cutting the adductor muscle that holds the shells so firmly together.

Starting to carve out the flesh.

Not all oystercatchers feed along the shore when the chicks are born. Some inland-nesting birds probe locally for earthworms, caterpillars, leatherjackets and other soil organisms. This food is brought to the chicks—which remain in the nesting territories. When these chicks grow up they, too, will be inland feeders.

In order to make a living most oystercatchers do not rely on probing skills alone. Worms, whether from land or sea, form only a part of their diet. A large proportion of what they eat consists of clams, cockles and mussels, and their ability to open and prepare this food offers them an alternative diet of great value—more especially so when hard ground makes probing impossible.

The skill shown by oystercatchers when feeding on mussels is almost unbelievable. They obtain the mussel flesh in two ways:

1. By 'stabbing'.
2. By 'hammering'.

It has been found that a bird will use one method or the other: it will be a 'stabber' or a 'hammerer'—but not both.

When lying underwater a mussel has its shells slightly apart so that it can filter from the sea the plankton on which it lives. And the 'stabbing' oystercatcher takes advantage of this by *stalking* mussels when they are lying open in shallow pools on the mussel beds.

First, the stalking oystercatcher thrusts its long bill between a mussel's opened shells and cuts the strong adductor muscle.

Then, having carried the now helpless bivalve ashore, the bird levers the shells apart by quick sideways movements of the bill.

Next, with *opening* movements of the bill it prizes the shells even wider apart.

Finally, it chisels out the flesh, using its bill like a pair of clippers.

Some oystercatchers, however, ignore the underwater mussel, preferring to feed on the exposed beds where the mussels are tightly shut. These birds are the 'hammerers'.

To open a closed mussel with only our fingers to help us is a very difficult task. Unless we have exceptionally strong hands we need to slide a knife blade between the shells and cut the adductor before we can prize the mussel open. The oystercatcher, however, can accomplish this feat using only its bill—one of the most astonishing feeding skills in the whole range of natural history.

Having torn a mussel off the bed, and then positioned it with the weakest part of the shell uppermost, the 'hammerer' uses its bill like a pick-axe to knock a hole in it.

Then, thrusting its bill through this hole in the shell it cuts the adductor.

After this, the shells are first levered open and then prized wider apart; the flesh is chiselled out into a single blob—and swallowed.

It is a very complicated operation, entailing as it does five separate skills: hammering, cutting, levering, prizing and chiselling. But the oystercatcher can do it with amazing speed. A mussel two inches long can be hammered open and cleaned out in under thirty seconds.

Research biologists have found that chicks reared by 'stabbers' become 'stabbers' themselves, and that chicks reared by 'hammerers' become 'hammerers'. By exchanging the eggs laid by 'hammerers' and 'stabbers' experimentally, it has been observed that the young birds develop the techniques of their foster parents. Thus, a chick reared by 'hammerers' becomes a 'hammerer', even though it came from an egg laid by a 'stabber', and *vice versa*.

So—these feeding skills are *learnt* by following the example of the parents. A rare case of learning by *watching*.

This passing-on to the young of specialized foods and feeding techniques could help towards the success of the oystercatcher as a species. It means that if a parent bird discovered a new source of food, or developed a new technique, it could pass it on to the young birds—and thus to the whole oystercatcher population.

The observation of colour-ringed birds has shown that these feeding skills take a very long time to perfect. Even though one-year-old birds can fend for themselves, they are still clumsy and inefficient. This has a most interesting consequence, as was found by systematically weighing birds of different ages throughout the year.

One shell has been cleaned . . .

And now the other—leaving a blob of mussel flesh on the tip of the knife blade. The severed adductor muscle (arrowed) can be plainly seen.

Oystercatcher's nest on shingle; the barest of scrapes, with a few pieces of flotsam. The camouflage of the eggs is marvellous. This picture has been taken in close-up so that they can be distinguished from the surrounding stones. To any but the keenest eye, the nest would be invisible from a range of more than a few yards.

Adults, who lose weight while raising young—due to their hard work on the feeding grounds—rebuild their body weight during the winter months.

One-year-old birds grow in autumn, but lose weight in the course of the winter.

Two-year-olds do better, but don't reach the weight required for successful breeding.

Even three-year-olds don't quite reach it.

This is why oystercatchers do not mate until they are four years old, they simply could not cope with the rigours of raising a

An oystercatcher chick stands gorged with mussel flesh provided by a parent. As the nature detective will be quick to spot, these mussels have been stabbed while lying open underwater in some sea pool—no holes have been hammered in the shells.

Oystercatchers enjoy a long life. One ringed bird is known to have reached an age of twenty-nine years—remarkable success for a ground-nesting species. They are usually four years old when they first mate. This seems a long time. But not until the young birds have had several years of practice can they open shellfish fast enough to feed a family as well as themselves.

brood. Being very long-lived birds, it is obviously of great advantage to wait until they are thoroughly efficient food-gatherers if as many young as possible are to be raised in a lifetime. Attempts to breed prematurely would result in weakening the birds, and thus jeopardizing the future.

Here is a real test for the nature detective.

The picture shows part of an estuary mudbank photographed at low tide. Redshank prints surround a number of little holes that have been made by a marine animal. But *what* animal? Is there a species of worm, or a clam, buried underneath each hole?

No! The curving line of holes is the feeding trail left by a flounder that has hunted across the mudflats at high tide.

Story between the tides

To set out across an estuary mudflat at low tide to track a fish seems a most unlikely enterprise. And yet it can be done in the case of the flounder, for this estuarine flatfish leaves an underwater trail in the mud as clear and distinctive as any animal on land.

The flounder preys on a number of tiny animal species that live in the surface layers of mudflats. It feeds by sucking in a quantity of mud and sand which is sifted through the fish's gill rakers to extract the food it contains. The sifted mud is then expelled

Above
A flounder almost completely buried in the sand. Only the eyes, mouth and branchial openings are visible.

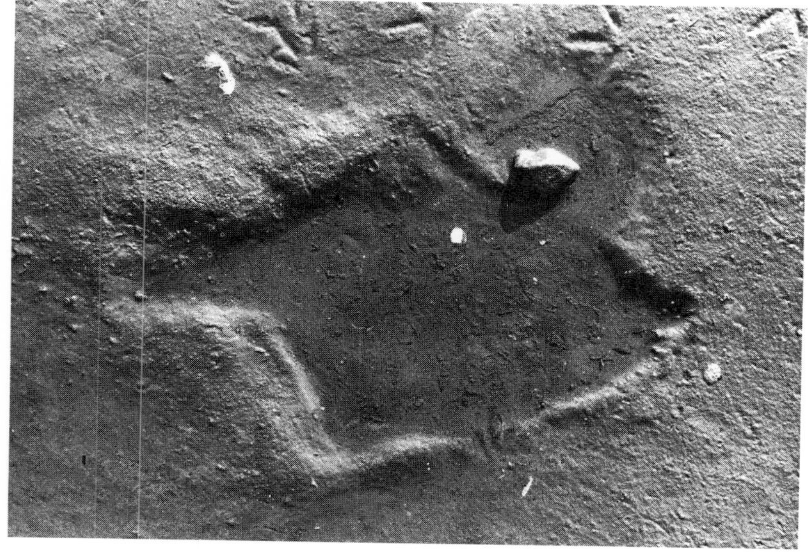

This flounder-shaped trough, photographed at low tide, shows where a flounder has been buried for a time during the period of highwater. On the floor of the trough are the sand tubes of spionids. Redshank prints can be seen nearby.

Postures adopted by the flounder as it makes a feeding hole and filters its food from the mud.

1. Supported by its fins, the fish shuffles across a stretch of tidal flats on the hunt for food.

2. At a chosen spot it tilts sharply upwards.

3. Then, arcing its body, it sucks in a mouthful of mud.

4. The food items in the mouthful are extracted by the flounder's gill rakers and the sifted mud blown out through the lower branchial aperture. After which the fish moves on towards another snack, leaving a trail of feeding holes behind it.

The dorsal and anal fin marks made by a flounder as it shuffled across the mudflats. Feeding holes can be seen at the start of the trail.

The feeding holes and fin marks of a flounder. Note the pile of sand that has been sucked in by the fish, and then blown out again through the lower branchial aperture.

through the fish's branchial aperture, and the flounder shuffles on across the flats to take a fresh bite—thus leaving a trail of feeding holes where the bites of mud have been taken out.

The flounder's diet varies. On the estuary flats of the River Ythan, in Aberdeenshire, where these pictures were taken, *Nereis diversicolour, Hydrobia ulvae* and *Macoma baltica* are all eaten, with *Corophium volutator* as the main food item. Although this diet is shared by many other predators, such as shelducks, sand gobies, oystercatchers, redshanks, dunlin and other waders, the flounder is by far the most important predator in the estuarine community.

The feeding times of the flounder are mainly during the periods of high tide on the mudflats, irrespective of whether high tide occurs by day or night. During the periods of low water the flounder does little feeding, tending to lie at rest on the bottom of the main channel.

A small area of some intertidal mudflats showing the high density of flounder feeding-holes. A hole remains visible for about four days, before being silted up.

What animals have left these tracks? Penguins? Marine crocodiles? Sea lions?

Leaving trails like those of some prehistoric amphibian, two seals have flipped their way across the sandy beach and into the sea.

A regular beachcomber

A creature usually associated with fields and woods but which is a regular and highly successful beachcomber is the carrion crow.

Hunting along the seashore, crows are quickly on the spot after a predator has made a killing—as can be clearly seen when we find their long, dragging tracks surrounding the corpses of gulls killed by foxes on the tideline. But carrion crows are not only eaters of carrion, they are always ready to hunt on their own account—in particular for the eggs and chicks of ground-nesting birds. An egg may be carried some distance from the nesting ground and buried under a tussock of grass. Or it may be hammered open and eaten on the spot.

Compared with other animals, crows seem highly intelligent. Most pronounced is their intense curiosity; their urge to investigate every new object they come across. Secondly, the speed with which they learn to associate certain of these objects with food.

An example of this was studied on the sands at Ravenglass after a westerly gale had washed up thousands of living razor-clams. Within a few hours the local crows discovered that this type of shell represented a tasty meal, and they had a wonderful time gorging themselves. For weeks afterwards they went round

Carrion crow tracks with regurgitated pellet. The crow's ability to exploit unexpected food sources shows it to be an accomplished nature detective.

Caporal, the young crow on my shoulder, was tamed by the scientists of Niko Tinbergen's research team on the dunes while *The Sign Readers* was being filmed. Like all corvids, crows are very acquisitive birds, and Caporal was no exception. Cigarettes, matches, tea spoons, pens, pencils and metal bottle-tops were among a host of items stolen and hidden. He even got away with a £5 note! Its owner, preparing a shopping list in the caravan that formed the team's laboratory, kitchen and office, put it on the table for an unguarded moment . . . Reminded by a flap of wings, he turned in time to see Caporal disappearing over a dune edge with the money in his beak.

Five minutes later the crow returned 'empty-handed'. A piece of scrap paper was offered to him in the hope that he would fly off and bury it in the same place. But of course he wouldn't—and despite an intensive search, the cache was never found.

hopefully turning over every razor-shell they came across.

As an experiment to find out how quickly crows learn to associate food with such hard, inedible objects as pieces of shell, a number of empty mussel shells were laid out along a stretch of sandy beach that was regularly visited by crows. A lump of meat was placed beside each mussel shell.

In a very short time the crows discovered the bits of meat and gobbled them up.

Next, to find out whether this single experiment was sufficient to teach the crows that mussel shells meant food, bits of meat were buried underneath each shell.

The crows came along, explored every shell and dug up the meat. They had learned very quickly. And although cockle shells were then mixed with the mussel shells in equal numbers, the crows still concentrated mainly on the mussels.

But later, food was hidden underneath the cockle shells. As soon as the crows found it they switched their attention completely. From then on the mussel shells were ignored.

A simple experiment, showing not only the speed at which crows can learn, but that like some other animals crows can read simple signs in much the same way that we can. Man is by no means the only nature detective.

Strange story of a lumpsucker

Near the low-water mark, beside a shallow pool among the rocks, lies a dead partly-eaten lumpsucker. This is a common enough species, and yet such a simple sign as a dead fish can tell an involved and interesting story.

Surrounding footprints show where herring gulls have fought over the carcase. Already, half the fish's head has been pecked away. The body, however, is intact. The fish is obviously a *male* because of its beautiful red belly, which the male alone develops during the breeding season—with the dual purpose of attracting a mate and repelling other males.

But how does it come to be here?

Right
Stranded lumpsucker. A favourite food of the grey seal, the lumpsucker is also eaten by sperm whales in Icelandic waters. Outside the breeding season it is usually found at a depth of about thirty fathoms. Females are larger than the males, growing to a length of two feet.

Below
'The scavenging gulls.' A typical scene: herring gulls feasting on the left-overs of modern society. Herring gulls and lesser black-backs are steadily increasing in number—the direct result of an affluent and expanding human population.

The southern tip of Walney Island off the north-west coast of Britain, close to the shipyards of Barrow-in-Furness.

The nature detective never knows when he may find something new. At low tide this great sweep of sands seems empty and deserted, and yet it holds a story of animal behaviour that until a short time ago was quite unknown: a story discovered entirely by the reading of tracks and signs.

Before reading the caption below, what do you make of these curious tracks in the tide-rippled sand of Walney Island? How many animals have been involved?

The tracks seem to have been made by an army of tiny creatures. But in reality only *one* creature was responsible. Buried underneath the little mound of sand (arrowed) at the end of its trail of claw-marks is a small edible crab. And what makes this such a remarkable picture is the fact that the edible crab is a deepwater species and, until the discovery, was not known to come up and bury itself above the low-tide mark.

Underneath its throat is a large and powerful suction disc with which it clings to a rock—hence its name. It seems an unlikely fish for herring gulls to be able to capture. But as their tracks prove, the gulls have found the fish in the pool and dragged it out on to the beach; and from its freshness it seems likely to have been alive when caught. But lumpsuckers live deep down in the sea; far too deep to be trapped in this way. What was a seemingly healthy male lumpsucker doing in such shallow water?

The answer lies in the extraordinary behaviour of these fish during the spawning season. Some time between February and May, on an inshore breeding migration, the female lumpsucker escorted by the male, swims into shallow water and lays her eggs very close to the low tide mark. After the spawning is over, the female returns to deep water, but the male fish stays behind to guard the eggs—fanning them with his pectoral fins and blowing a current of water over them with his mouth. Whatever happens, he will remain with the eggs until hatching: a period of six or seven weeks. So that on an exceptional spring tide, when the sea ebbs to its maximum distance at low water—as it has on this occasion—the egg-guarding male lumpsucker may easily find himself stranded in some shallow pool. And then, of course, he becomes an easy prey for the scavenging gulls.

So—the fact that the stranded fish is a *male* and not a female is exactly what the nature detective might expect.

Sign of Cancer

The discovery of the buried crab was made quite by accident, when at low tide a slight dome in the sand was noticed close to the water's edge. The finder, a schoolgirl, kicked the dome and found to her surprise not the pebble that she expected, but an edible crab.

Soon, other domes were found and although some of these *were* caused by buried stones, many contained crabs. Later, it was discovered by Niko Tinbergen's research students who were studying the gulls on Walney that when the beach was covered with the tide during spring and early summer, tens of thousands of young edible crabs came inshore and dug themselves into the sand, remaining there when the tide ebbed, their presence betrayed only by the little sand domes.

But further tracking revealed something of even greater interest. Some animal had already solved this little riddle of the sands.

The herring gull.

Crab remains, scattered sand and gull footprints showed where gulls had hooked crabs out of their holes and eaten them.

Observation soon established that the gulls didn't stumble on crabs by accident. As the tide ebbed they flew out from their nearby nesting ground and hunted purposefully along the shore, stalking about and pecking at every dome they came across.

This posed a problem. How did a gull *know* there was a crab underneath a dome? Was it scent, the sight of the crab, or the dome itself that attracted a gull?

Further tracking provided the answer. Places were found where a gull had walked along and pecked at what *looked* like a crab dome, but was in fact a bubble caused by air trapped under the sand. A number of pecked air bubbles were found, proving that (since there were no crabs underneath) it could not have been the scent or sight of a crab that had attracted the gull—it *must* have been the dome itself.

So—simple nature detective work, starting with the casual observation of a schoolgirl, had resulted in two fascinating discoveries: that young edible crabs will sometimes leave the deeper water and spend a part of their time ashore; and that the herring gull is itself a 'nature detective' and able to hunt by reading signs.

Tens of thousands of herring gulls and lesser black-backs—among whom these jackdaws are unconcernedly feeding—nest on Walney Island each year. When the breeding season is at its height, many tons of food a day are needed to support the huge colony of birds—whose numbers have been multiplied by the voracious and rapidly-growing young. Much of this food comes from the nearby coastline where the parent gulls—more especially the herring gulls—forage daily for what the sea provides. It is the beachcombing herring gulls that have discovered the unexpected source of food hidden in the sand.

Bird city

Nature detective work does not, of course, stop at the elucidation of footprints and signs, but extends across the whole range of animal behaviour. To a large extent, an understanding of animals depends on our ability to interpret the way in which members of a species communicate with one another.

Animals of the same species understand each other's intentions, just as human beings do. Each species has its own 'language', which enables individuals to communicate with one another and achieve the co-operation that leads to successful breeding.

In the picture below, the young lesser black-backed gull (on the right) is pestering its parent for food, signalling its hunger by pecking at the base of the parent's bill. At the same time, it adopts a hunched posture, which signals: 'I mean no harm'.

The parent bird understands both these signals and, by its contortions, shows a curiosly ambivalent attitude. It cannot resist the young bird's demand for food; but, at the same time, tries to avoid being pecked-at and overrun. Were it not for its signals of appeasement, the young gull would probably be driven away.

Nearly all the sounds an animal makes, and many of its postures and movements, together with some of its coloration, are signals to others of its own kind: mainly, signals of aggression, appeasement, danger and love. Our understanding of the different 'languages' used by various species, demands considerable feats of observation and deduction. But when we begin to grasp the meaning of what animals are doing, much of their everyday behaviour—which, hitherto, we may have taken for granted—assumes a new significance and a new interest. A wonderful opportunity for some original discoveries by the observant nature detective.

A mass of small shore crabs litter this herring gull nest site. An example of selective feeding.

The lesser black-back tends to do rather more fishing out at sea than the herring gull (in spite of the latter's name). But both species range along the tideline—and scavenge on refuse dumps.

Dismembered starfish remains twenty yards above the high tide line. A herring gull's work.

The study of animal communication is like learning a series of new languages, and it has a compelling fascination. A superb example of how a system of communication 'works' can be seen in the social behaviour of a colony of herring gulls and lesser black-backs, such as the one on Walney Island not far from my home.

A good deal is known about these two very closely related species thanks to Niko Tinbergen's research—on which he and I based our film: *Signals for Survival*. As I soon discovered when we started filming, detective work in the field such as studying the behaviour within a gull colony can be highly rewarding, both for the drama and the humour it reveals. But anyone who seeks to understand the gulls' community life must begin to watch as soon as the birds arrive. For although no eggs will be laid for another four to six weeks after this, these early weeks are a period of intense activity and an understanding of what is going on then is essential if an observer is fully to appreciate the birds' social relationships during the rest of the season.

All through the winter, like those of the black-headed gull, the big gulls' breeding grounds are deserted. During late March, or early April, after some days spent in reconnaissance flights, the flocks of herring gulls and lesser black-backs will take possession of the nesting area.

Anyone who had not before observed such a sight could be forgiven for thinking that the great mass of screaming gulls was totally without any form of order or discipline.

The mewing lesser black-backed gull on the right leads its mate towards a possible nest site. The 'mew' call, which in gull language means 'follow me', is a very important signal. That the birds should agree on the position of the nest is essential; for, when the eggs are laid, both male and female will take it in turn to incubate.

Left
While still inside an egg, a chick is in touch with its parents, by sound. It understands parental calls even before it has hatched.

Above
Instinctively, a chick understands certain signals and can communicate with its parents. The dark red spot near the tip of the parent bird's lower mandible is a colour signal: it stimulates a chick to peck at it when hungry. The parent in turn responds to this pecking by regurgitating food and feeding the chick. If the chick failed to signal that it was hungry, it would starve to death.

But although at first sight a colony of herring gulls and lesser black-backs seems chaotic, an observer, who has the patience to wait and watch unseen, will soon realize that this is no haphazard gathering but is, in fact, a great bird 'city'; a stable social structure extremely well organized by its inhabitants.

Nevertheless, however good its organization, this teeming concourse of gulls is by no means a city of friends. Far from it. Each bird is a potential cannibal, ready to steal its neighbours' eggs; to kill and eat any neighbouring chicks. That such a huge number of cannibals can breed successfully when packed so tightly together, their nests only a few yards apart, seems astonishing. And yet succeed they do; the beautiful efficiency with which they manage to rear a family resulting from a highly developed gull 'language': a mutually understood system of communication that comprises posture, movement, sound and colour.

Inside the colony each pair of gulls occupies a territory: a few square yards of ground defended by its occupants and respected by everyone else. Just as we, in our cities, enjoy comparative safety within the sanctuary of our own houses, so the gulls of 'bird city' find comparative safety inside their homes—or territories—which are formed solely for the purpose of breeding.

The owner of a territory is always the male bird. Early in the breeding season he 'stakes his claim' to a piece of ground and proclaims his ownership by strutting about and shouting: 'This is *my* patch! Keep off!'

This message, delivered at frequent intervals to everybody within earshot, consists of a loud trumpeting cry known as the 'long call'. He achieves this by bending his head down almost between his legs, then throwing it up and uttering a prolonged bellow—'Whooooo . . . waugh, waugh, waugh, waugh, waugh, waugh . . . !' Coming from his great foghorn of a throat, this call can be heard for a considerable distance.

In defence of his territory, a male is prepared to fight any other male. When he does he uses his wings and his bill as weapons of attack. The bill is used for jabbing downwards. The folded wings form clubs for beating with.

But it is seldom that a male has to fight seriously for his territory. Seeing and understanding the other bird's posture of threat—the weapons (beak and wings) held at the ready—is usually enough to prevent bloodshed. When a fight does take place the outcome is seldom in doubt. The winner is the bird that is fighting on his own ground. Almost invariably the intruding gull is driven off.

That the territorial owner should always win seems remarkable. But, even more remarkable, an intruder usually runs away before he is attacked. The reason is that each male, when outside his own territory, is afraid of all other males. He will attack any male trespassing on his own territory, but will run away from that very same opponent if he finds himself on that opponent's territory.

During boundary clashes the birds are torn between aggression and fear. Each feels impelled to attack the other, but is afraid to cross the boundary line (thereby 'putting himself in the wrong') to do so. The result is that each bird stays where he is, with his weapons half-drawn in an attitude of threat.

This intimidating signal of 'threat' has a most useful consequence. It acts as a deterrent. And the human parallel is obvious.

Two angry male gulls, face to face and threatening each other over their common boundary, are like two men arguing over a garden fence. Each is waving his fist at the other, and longing to fight. But inhibited from doing so because he is afraid to cross the fence and put himself, so to speak, 'in the wrong'.

But such pent-up emotion *must* have an outlet. The birds simply have to 'let off steam'. And they do. They peck violently at the ground, tearing out beakfuls of grass.

This 'grass-pulling' is a movement very similar to pecking a hated rival, but aimed at something inanimate. In other words, it is a re-directed attack—just as we, when angry but afraid to fight, may thump the table instead of our opponent.

As a result of this deterrent, fighting is reduced to a minimum. Territories are established and defended with the minimum of bloodshed. Valuable time during the breeding season, spent gathering food for the young, is not wasted by constant territorial fighting. Largely by the language of threat, a breeding territory is secured in which the future family can grow up unmolested by its neighbours.

Beachcomber par excellence

Although each of the animals that hunt their living along the shore is an expert in its own right, the star beachcomber, the true professional, is the herring gull. A versatile bird, it seeks its food in many locations: fishing ports, inland fields, refuse dumps—and indeed it can earn its living in these places—but its natural hunting ground is the strip of shore between the tides.

Often keeping it company is its close relative the lesser black-backed gull. Their habits are very similar. But the lesser black-back is more of a fisherman and slightly less of a beachcomber than the herring gull—despite the latter's name.

The gulls know that the shore provides most food when the tide

Perfection in flight. Outlined against a stormy sky. A hunting gull ranges along the waveswept beaches, keen-eyed for a meal.

is ebbing, and they know when this will happen—even though it occurs at a different time each day and the birds are miles away, far out of sight of the sea. Gulls are not alone in possessing this strange knowledge. Most impressive are the beautifully timed curlew flights that arrive on the salt marshes almost to the minute as the mudflats uncover with the ebb tide, the birds having come from fields several miles inland. Crows and oystercatchers are two other species that share this proclivity. Quite how these birds can tell the time so accurately by their 'tidal clock' is a question that nobody can answer.

The gulls' search for food is very much conditioned by the weather, their hunting grounds being chosen accordingly. On calm days there is the chance of good fishing out at sea. Swirling like snowflakes against the dark water the gulls plunge-dive for young herrings that have been driven to the surface by pursuing mackerel. But in wild weather the best pickings are to be had along the shore.

As the tide ebbs, the birds go out with it to hunt the animals thrown up by the rough sea. Crabs and whelks, left stranded by the preceding waves, try desperately to dig into the sand, but the keen-eyed gulls soon hook them out. The gulls swoop on to every scrap of food, ever ready to rob each other. No meal is safe until it is swallowed! Fish, crabs, starfish, sea urchins, shellfish . . . all the accidental spoils of the sea that have led to the evolution of beachcombers who put all this food to good use.

Each beachcomber has its own method of feeding, and when it comes to dealing with shellfish the herring gull is no exception. An oystercatcher cracks a mussel open with its bill, levers the two halves of the shell apart, and chisels out the flesh. But the herring gull has never learned how to smash a mussel open with its bill—powerful though it is. Instead, the bird flies up with a shellfish . . . and drops it. If the shell happens to hit something hard—rock or shingle—all well and good, it is almost certain to crack. If not, if the dropping area is a patch of soft sand, the gull will have to try again and again.

After years of trial and error, the older gulls seem to learn that a shellfish has to be dropped on a hard surface; but the younger birds drop entirely at random, their prey falling time after time on soft sand—or into the water.

On stormy days, when a lot of food has been stranded, young gulls can often be seen dropping over and over again all along the beach—flying up at a very steep angle into the wind, and then suddenly tilting downwards. The nature detective will know exactly what is going on, even though the birds are a long way off. The pattern of flight is unmistakable.

The fringe of the sea: a constant source of food. Not only for the gulls, swooping up into the grey, stormy sky to drop their mussels or whelks, but all the other species that feed along the shore. Each is adapted to earn its living in its own way; to make use of a constant supply of discarded life that, far from being wasted, is the start of numerous food chains.

The masses of marine organisms washed in by every tide feed a myriad sand dwellers. Larger creatures eat the sand dwellers. And even the carcases of these larger creatures are not wasted. Dead animals, whatever their size, are food for a host of scavengers—from beach-hoppers to black-backs—and so remain an integral part of the life-cycle along the narrow strip of shore beside the sea whose discards feed so many.

Part 2

The Countryside

Although, as we have seen earlier in this book, many animals are active mostly at night, all of them leave tracks and signs from which, in daylight, they can be identified. Along a muddy lane while on a country walk, we may find badger or rat or fox tracks. In the woods, fresh scars on tree trunks, where the bark has been peeled off by feeding red deer. A ravaged sapling with earth pawed bare at its base, typical of the roe buck's territorial marking. In a new plantation of young trees, bitten leaders show where hares or rabbits have been at work. Soft earth on the margin of river or pond may hold prints of heron, wild duck, moorhen or otter; or the probing holes of snipe. Pellets disgorged by owl, buzzard, rook and many other species, divulge how these birds have been getting a living.

Some signs are very simple: wormcasts, mole hills, rabbit burrows, woodpecker holes, a rookery. Others are equally unmistakable though less numerous: a thrush's anvil; the plucking-blocks of various raptors; a shrike's larder—perhaps a bumble-bee impaled on a barbed wire fence; a pheasant's dust bath; or, deep in the woods, a badger's sett with its mounds of earthworks.

Many species can be identified from their droppings: pheasant, grouse, wild goose, deer, hedgehog, badger, fox, to name only a few. And there are loose feathers, scratches and scrapes, caches of nibbled nut-shells, stripped pine-cones, tufts of fur and thousands of other clues left by the busy animals of the countryside.

Observing these clues and building up stories from them can

Opposite
A glorious sweep of Cumbrian countryside. Our land has for ever been in a state of change. These fellsides were once densely forested, mainly with oak and birch—cleared centuries ago for charcoal and to provide grazing land for sheep, whose appetites prevented the natural regeneration of trees.

But apart from a narrow cultivated strip beside the river, and an outline of bare fells (which might be considered preferable to a line of factory chimneys) the valley is much as it was when Agricola's mercenaries marched past this very spot.

add enormously to our enjoyment of the natural world about us. But if we really wish to understand the signs that animals leave, we should first make sure that we know the animals themselves and at least the rudiments of their behaviour; for it is impossible even to guess at the origin of a footprint, scrape, feather, dropping, pellet, or any other sign, if we are ignorant of the species that left it.

Before setting out to track in any chosen location, it is sensible to ascertain what animals are likely to be found there, at what time of year, and why. Gradually, as we gain experience and expertise, our detective work in the field will confirm (or confound) this, and perhaps lead to some surprising discoveries. As we have already seen, a number of old friends keep turning up in unexpected places, showing some quite startling traits. And where nature detective work is concerned, it is important to remember that a keen-eyed schoolchild is fully capable of making some fresh discovery.

Hen pheasant's dust-bath in dried soil on the edge of a woodland ride. There is neither snow nor soft mud to hold tell-tale footprints in the heat of mid-summer, but the dusting bird's sex and identity are betrayed by the feathers (arrowed).

Pheasants were introduced to Britain sometime towards the end of the Roman occupation. The Romans certainly knew a good deal about rearing them. Pallacius, in *De Re Rustica*, gives detailed instructions.

Animal signatures.

The person who carved his name in the bark of this tree made no secret of his identity. But man is not alone in signing trees. Many animals use trees in one way or another and can be just as easily identified by the signatures they leave behind them.

Below right and below
Hazelnut pushed into crevice in tree bark by nuthatch.

'The difference between the methods employed by the *nuthatch* (*Sitta europaea*) and the *nutcracker* (*Nucifraga caryocatactes*) is neatly shown by their names. The *nuthatch* is the 'nut-hacker', the provincial 'nut-jobber', and is so called from its habit of hammering with its bill until it splits the nut which it has previously, if necessary, wedged in a crevice in a tree. The *nutcracker* (a modern formation on *nut* and *crack*) was noted by Turner as *nucifraga* (L. *nux*, 'a nut' + *frangere*, 'to break') from the dialect name of a bird which he saw in the Rhaetian Alps.'

Stephen Potter and Laurens Sargent, *Pedigree: Words from Nature*.

Above
Woodpecker feeding-holes pit the rotting stump of this pine, which has been killed by fungus caused by the depredations of a much larger forest animal (see p. 206).

Below left
Britain has three species of woodpeckers: the Green Woodpecker or 'Yaffle'; the Great Spotted Woodpecker and, more locally, the Lesser Spotted. All three species feed on the larvae of wood-boring insects, spiders, seeds and berries; but the Great Spotted variety (that has been pecking for grubs in this rotting stump) is also known to take the chicks of small birds.

Below right
The 'signature' of a green woodpecker or yaffle. The yaffle's undulating flight and loud, clear, rapidly repeated call—sounding like a derisive cackle of laughter—is one of the most distinctive features of the countryside.

Above
Abandoned woodpecker holes provide nesting sites for many other species . . .

Above right
A 'den' tree. Home of many different species. Green woodpeckers and great spotted woodpeckers nest here, and in some of their abandoned holes redstarts, starlings, titmice, pied flycatchers and other birds have reared their families. Insect-eating species are the forester's friends, for some insects can cause great damage to trees. But not all holes are as safe as they look . . .

The upper hole is the entrance to what has been a coal tit's nest of young. Safe enough in there, one might think. But shortly after the chicks had hatched, a great spotted woodpecker knocked its way in through the lower hole and stole the young.

Mr M. C. Wilkes, who took the photograph, writes: 'This was of special interest to me because I had never seen it happen before . . . The nest of young had just hatched out and I was all set to photograph them. The following day I found the second hole had been made and all the young had gone . . . A very insidious practice of the Great Spotted Woodpecker.'

Opposite
The great spotted woodpecker advertises its presence by a sort of mechanical drumming—a loud, vibrating sound caused by a rapid rain of pecks on favoured dead tree branches.

Nesting holes of a sandmartin colony. Snug inside their little holes set in the sheer cliff face the birds are safe enough. But no small bird is safe while on the wing. Many remains of a flier who 'failed to return' will be found on a hawk's dining table . . .

Birds taken by hawks are carefully plucked before being eaten, each species of hawk doing so in its own characteristic way. From a study of the 'kill', an experienced observer can identify the killer from the plucking technique.

Meal remains disclose the presence of many species of animals, some of which are seldom seen in the wild. They also provide a glimpse of animal feeding techniques.

Plucking-block, where goshawk has plucked a woodpigeon.

The feathered 'gun'

It is unlikely that either sling or bow was used much for shooting birds in flight. The substitute for the gun was the hawk. Hawking is of ancient origin, being developed it seems in early Thrace. It was certainly practised in England in Saxon times, and probably much earlier.

Once, all persons according to their rank were assigned their own particular species of hawk, the more noble species (often imported at great cost) being reserved for royalty. The falconry section of *The Boke of St Albans*, attributed to Dame Juliana

A sparrow-hawk's plucking-block.

Berners, Prioress of Sopwell Priory, describes the 'order of merit' in fashion at the time. First was the *gyr-falcon*, Chaucer's '... gentyl faucoun that with his feet distrayneth The Kynges hond.'

The *peregrine* was reserved for princes and higher nobility. The *saker* was for the knight. For the squire, the *lanner*. For the lady, the *merlin*. For the 'young man', the *hobby*. The *goshawk* was the yeoman's bird, whereas the tiercel, her mate, was the 'poor man's' hawk. Priests and clerks in minor orders flew *sparrow-hawks*. And, lowest of all, the *kestrel* was for the 'knave' or servitor.

Well trained birds exchanged hands for huge sums of money. In the reign of James I, Sir Thomas Monson paid £1,000 for a pair of goshawks. And one of the crusader princes is said to have been ransomed from the Saracens for twelve Greenland falcons.

Today in Britain, all hawks are protected by law, and this legislation is by no means new. Although protection lapsed in the 18th and 19th centuries, earlier laws were very stringent. Eyries and nests were protected, and anyone who stole another man's hawk was heavily fined or imprisoned.

Sparrow-hawk taking a bath. It usually does so after a meal. According to M. H. Woodford in *A Manual of Falconry*, 1960, the female hawk is more disposed to bathe than the male, which seems to be more afraid of being caught at a disadvantage with its feathers wet.

Squirrel's dining table.

The grey squirrel, a species introduced from North America, is far from popular on account of its destructive habits. Grey squirrels will eat almost anything from tree shoots to eggs and young birds.

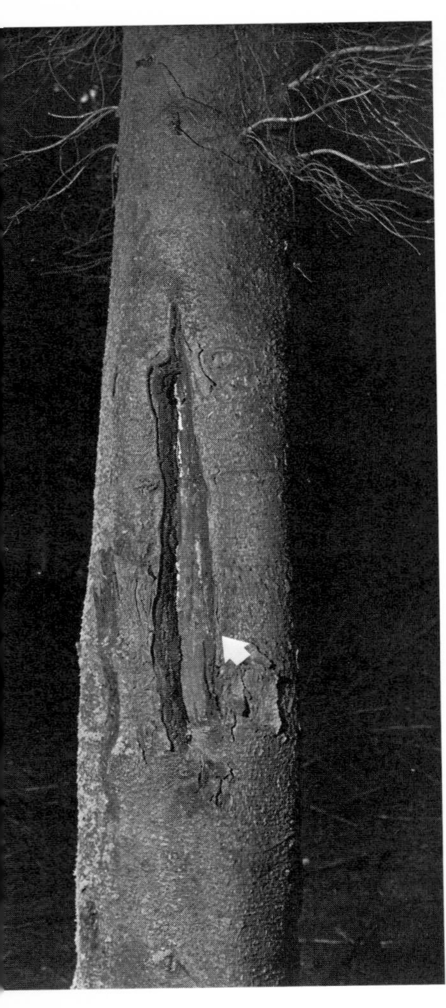

The scar on this tree trunk was caused neither by penknife nor beak, but by the teeth of a red deer, Britain's largest species of wild animal.

Cones of Norway Spruce stripped by red squirrels.

A song thrush cracks snails open by taking them in its bill and hammering them on a hard surface. Usually a stone, but here a piece of wood.

Above
Thrush's anvil surrounded by cracked snail shells. Note: Both stones have been in use.

Top right
Hazel nuts, chiggled by woodmouse (or long-tailed fieldmouse).

Bottom right
Woodmouse work again. This time, sloe. Both sets were taken from a large cache of nuts found in a mossy crevice between rocks underneath a blackthorn near to my cottage.
The nut-work of woodmice can be distinguished from that of bank voles by the toothmarks. As can be seen in the pictures above, woodmouse chiggling shows radiating toothmarks on the inside surface of the hole, with irregular marks left by upper incisors around the outside edge. Nuts opened by bank voles show no toothmarks visible on the outer surface of the shell.

This toadstool, the stinkhorn (*Phallus impudicus*), is covered with bluebottles. Why?

A nature detective might infer that the stinkhorn has some insect-attracting property that is designed to do it a bit of good.

He would be right.

The stinkhorn exudes a glutinous spore-mass smelling of rotten meat. Spores stick to the feet of insects attracted to the 'feast' and get planted out when the insects fly off and land elsewhere.

Common or garden.
The grey squirrel is a familiar figure in parks and gardens, but the fieldfare, below, is lesser known. A winter visitor, it sometimes visits our orchards to feed on fallen fruit.

Age-old creatures of the night

Badgers have lived in Britain for thousands of years, but with the increase of urban encroachment are having to adapt to rapidly changing conditions. This they can do very well, if left in peace, for they are truly omnivorous and will eat an astonishing variety of food. A favourite dish is the earthworm, but they are also very fond of young rabbits and rats, mice, voles, eggs and chicks. They will glean in the cornfields, eat hedgehogs, slugs and large numbers of insects, munch up any fruit they come across and dig for bulbs and roots. They love wasp grubs, and have even been known to tackle hornet's nests.

From illustrations in the popular press, the covers of many nature books, and even on postage stamps, the characteristic striped head has become a very familiar sight. But the badger is a creature of darkness and, apart from a fleeting glimpse in their car's headlights along some country road at night, few people have ever seen a badger in the wild.

Signs of badger feeding activity. Disturbed grass and soil show where the animal has been nosing out roots.

The badger has poor sight, but keen hearing and well-developed power of scent. Its tough snout is used as a shovel for digging up food lying close to the surface (see picture opposite).

Nevertheless, provided they make adequate preparation and use great stealth, keen nature detectives need not despair of watching wild badgers. Having satisfied themselves that a particular sett is occupied, and made a thorough study of the surroundings, they may well see something of badger behaviour during the hour or so before dark. It is, however, essential to be *quiet*, and never to be upwind of the sett, for badgers have exceedingly sharp ears and the keenest sense of smell.

The use of a hide is recommended, or perhaps a perch in the lower branches of a convenient tree. Although, far better than allowing oneself to be silhouetted against the evening sky is to use natural ground cover at least twenty yards from the nearest entrance in use, or even to stand motionless against a tree trunk.

Badgers will desert a sett if unduly disturbed. Only if they can watch badgers without the badgers knowing they have been

Entrance to a badger sett. The fresh-trodden earth and absence of a spider's web across the mouth of the hole indicate that the sett is in regular use. Other signs that confirm the presence of badgers are: footprints; newly-dug lavatories, and claw marks on the boles of nearby trees—unmistakable badger 'signatures'. Also, since the badger is a clean animal and seldom fouls its nest, there should be traces of bedding that has been brought out to air—and later dragged back again.

Hanging across the river not far from my cottage is this small suspension bridge. During a recent summer, a badger came from its home deep in the wooded fellside and occupied temporary quarters in an old rabbit burrow beside the river, sleeping during the day and hunting along both banks of the river at night. It crossed over by the bridge and dug itself a lavatory alongside each end—handy for whichever side of the river it happened to be hunting on.

Another entrance to the same complex of tunnels. A well-worn pathway stretches away to the right (arrowed). Badgers have lived in our woods since long before Britain became an island. Some may be using the same tunnel systems that were mentioned in the Domesday Book.

watched can observers consider themselves true naturalists.

Like many other animals, badgers have their regular trods. These are used whenever the animals set out on foraging expeditions or travel from sett to sett. So well-worn and distinct are these trods that in ancient times, when Britain was densely forested, the age-old network of badger pathways may have been the origin of some of our country lanes.

Badgers are very set in their habits. If a rabbit fence happens to be erected across their pathway, they will break straight through it. Weighing between 25 lb and 30 lb when fully grown, they are powerful animals. Some over 40 lb have been recorded. The woodsman, who wishes no harm to the badgers but is anxious to protect his young trees from being eaten by hares and rabbits, takes advantage of the badgers' force of habit by building special swinging gates into his fence at points where it crosses the badgers' trod. The powerful badgers soon learn to push open the gates—which are too heavy to be opened by rabbits or hares.

During summer and autumn, badgers move about with less

Badgers at night, close to the sett.

caution than in the spring, when cubs are being cared for, or in the winter when they live to a large extent on their accumulated fat and feeding excursions are less frequent. The mating of badgers can take place during any month between March and October, but irrespective of when she is mated, a sow badger always has her cubs in early spring.

Towards the end of the summer, neighbouring families tend to move into large social setts. Thirteen badgers have been seen to emerge from the same sett on an August night.

Badgers do not really hibernate. Even in snow there is usually some sign of activity near to the sett. Deep down underground the sett in winter is warmed by the heat of many badger bodies, so that on a frosty morning condensed water vapour steams from the mouth of the hole—an obvious sign of occupation.

More tree damage. What animal has been responsible for *this*?

The great bark theft

A detective story in four pictures.

This long branch (arrowed) shows up white against the background. It has been almost completely stripped of bark. A grave theft. Who could have done such a thing?

Close-up shot of branch pictured above. The bark seems to have been *nibbled* off. Who was the nibbler? Are there no clues?

Indeed there are. These light-coloured hairs identify the culprit . . .

And here he is!

Signs that tell the nature detective an interesting story: the work of leaf-cutter bees on rose leaves (subsequently pressed and dried). Leaf-cutter bees use pieces of leaf (usually rose or lilac) to construct the cigar-shaped cells in which their eggs are laid. The nest—which is usually made inside a hole in the ground, or in a hole bored in a wooden post—contains a large number of these cells. The longer pieces of leaf are used for the cell bodies; the circular pieces for the cell tops and bottoms.

When the young bees hatch they emerge from the 'comb' in reverse order. The first bee to eat its way out will come from the last cell to be completed. The rest follow suit. In this way each bee has only one layer of cell covering to bite through.

Country-garden magic

Over 20,000 animal species exist in Britain, each with its own way of life, and the nature detective need not travel far in search of stories that capture the imagination and fill the mind with wonder. Even a rose leaf in a cottage garden may hold its own tiny piece of animal magic.

In the complex jig-saw pattern of nature, each living creature depends upon another; each environment produces its own creatures, and each creature its own unique behaviour.

Sunshine glancing through a tracery of garden leaves lights up a tiny corner of the world of plants, vital to this living pattern. The sun, source of energy for all living things, activates the plants. Plants feed on inorganic materials which then synthesize into living matter—the food, directly or indirectly, on which all animals depend. (An area lacking in plant food will produce no animal life whatever).

Most ponds attract numerous species of wildlife. A search along the edge of the water will soon show who has been visiting, and when.

Unless the ground is frozen, soil on the margin of ponds and lakes is usually soft enough to hold footprints, and even when these are absent, there are other clues . . .

Evidence of recent visitors: mallard feathers, floating at the water's edge.

Feathers lying on the ground are clues that tell a number of stories. A single feather, or just a few, will probably have been left by a preening bird.

A scattered ring of feathers are usually those plucked by a hawk.

A *trail* of feathers denotes a fox kill. *Not* badger. Badgers usually eat their kill on the spot. Very seldom do they carry it away.

Plant life absorbs carbon dioxide from the air, retains the carbon—the basis of organic substances—and returns oxygen. It is to the plants that we owe the free oxygen in the atmosphere which has been accumulating for aeons, and on which our lives depend. And beneath our feet, even in the little gardens outside our homes, is the world within the soil from which all plant life springs. A world of microscopic organisms, as vast and complex and varied and as rich in life as the world above it.

It is a constant cycle. The soil gives up its chemicals to plants, plants are eaten by animals; animals die and break down into chemicals, completing the living pattern, which achieves its direction from the process of natural selection. Whatever its size, habitat or behaviour, every creature plays its part. In the natural economy, nothing is wasted. Animals interact not only with themselves but with the plants, and inorganic elements—the soil and water and air. A worm comes to the surface of a lawn and drags a fallen leaf down into the earth, and the leaf fertilizes and enriches the soil from which the tree itself has grown.

The droppings of animals using a pond help to fertilize the water with minerals—which means more plant life, more snails and shrimps. In turn, more food for residents and visitors alike.

This tiny island has been well trampled by mallard and teal, their identity betrayed by the feathers they have left behind.

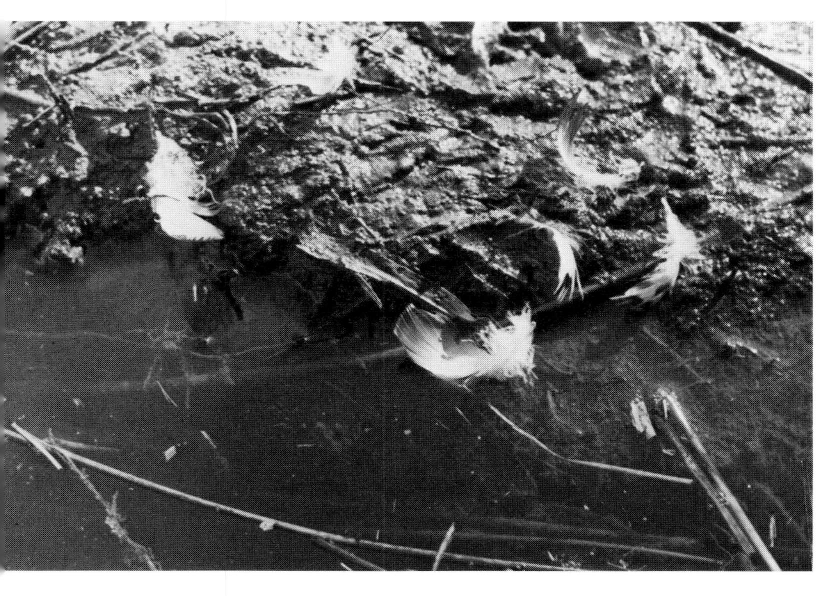

To repel water, a duck's feathers need frequent oiling. With its bill the bird squeezes oil from the oil gland near the tail, and then preens itself. Feathers quickly become water-logged when detached from a bird. When still crisp and curly, as these are, it is evidence that the bird has shed them very recently. Older feathers would be lying flat and 'lifeless', soggy with water.

Other birds have visited the waterside. Grey partridge, with (above right) pheasant.

Moorhen.

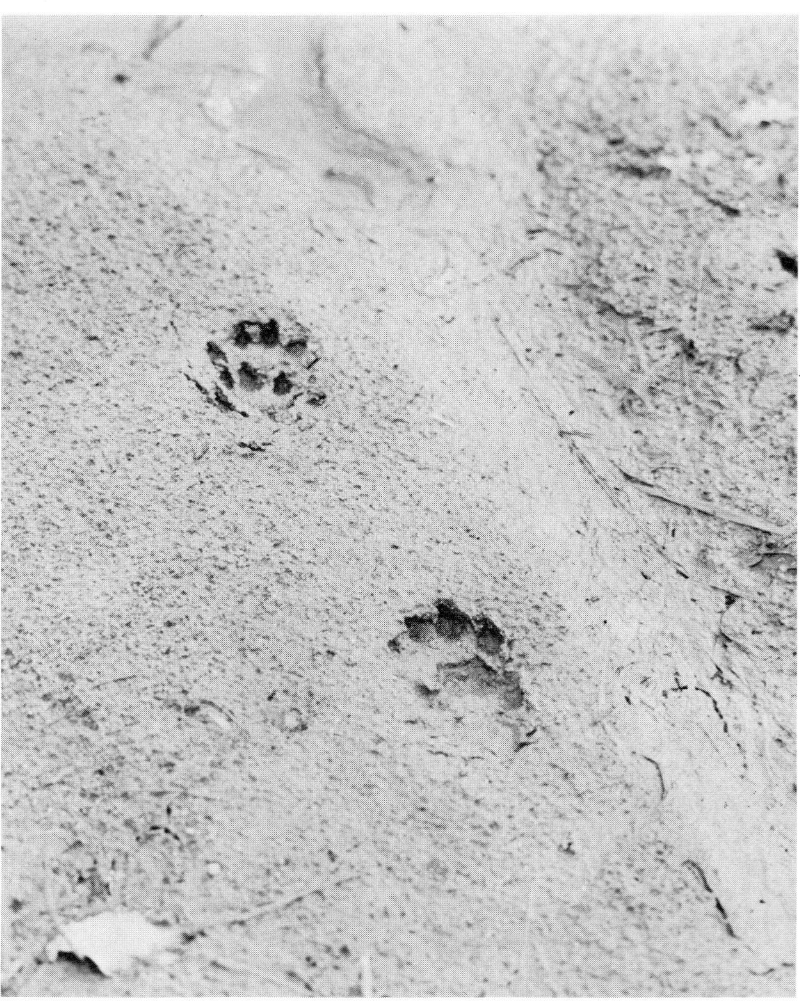

Wild cat prints found beside a freshwater loch in Argyllshire.

A pond, however small, is a focal point for wildlife. Sooner or later, during prolonged dry spells, most animals are attracted to the waterside.

Rare photograph of a nightingale in water. Mr M. C. Wilkes, who took the picture, comments as follows:

'This bird was heard singing in the corner of a wood where there was no water. I dug a hole approximately six feet square and one foot deep, lined it with polythene and filled it with water, then sat and waited for this shy bird to come down for a drink. After many hours it came down, and stayed long enough for me to get one shot in black and white.'

Underwater death-trap

To a knowledgeable observer on the bank, the surface of a pond can tell many stories of animal life and death. But there is none so astonishing and dramatic as that of the bladderwort. The distinctive yellow blossom of the bladderwort waves gently in the breeze, but to anyone who recognizes it, this beautiful flower tells an almost unbelievable story of the macabre.

The bladderwort depends on the flesh of water animals for its food supply. Although it grows in the open air, it sets its rows of animal traps underwater. Extending beneath the surface, the stem of the plant branches profusely, each branch carrying a number of tiny bladders that trap and digest tiny water creatures.

These bladders are simply rows and rows of little stomachs!

No more is known about the origin of this extraordinary plant than about the many other carnivorous species. How long it took to evolve, and what it evolved from, have yet to be discovered. But what *is* known about it is the deadly efficiency of its trapping system.

Trailing tendrils of the miraculous yellow-flowered death trap *Utricularia*, the bladderwort plant, common in northern hemisphere ponds and lakes showing some of the profusion of bladders in which the plant's prey are trapped.

Although no more than two millimetres in diameter the plant's bladders are lethal traps for aquatic animals. They can be round, triangular, cylindrical, pear-shaped, but they all have one feature in common—a small semi-circular trap-door operated by a cluster of tiny triggers set at the base of a funnel-shaped growth of guide hairs.

These guide-hairs funnel an approaching prey in towards the triggers that 'spring' the trap-door. Exactly how the triggers work—how they transmit the impulse to the trap-door hinge mechanism—is still in doubt. But there is no doubt about what *happens*.

When a trigger is touched, the trap-door flies open inwards. Water is sucked inside, and the prey is sucked in with it. Instantly, the trap is re-set, ready for another victim.

The bladder is in effect a 'blunder' trap. But there are plenty of animals to blunder into it.

An unpolluted pond holds a vast number of tiny creatures that can be caught by the bladderwort. Common among them are mosquito larvae and daphnia, or water-fleas. During the time of year when the bladderwort is growing bladders, daphnia multiply at a tremendous rate and form a large proportion of the plant's diet.

A daphnia or a mosquito larva may be trapped by just a single antenna. Even so, the animal is doomed. However hard it struggles it will never get free. Sooner or later some part of its body will touch the triggers again—and inside it will go, its body providing more nitrogen for the bladderwort plant.

All along the margin of a swamp, thousands of bladders are in action; lines and lines of traps clinging to their assortment of victims like rows of gallows.

Viewed from the pond-side, the bright yellow bladderwort waves in the sunlight. To the outside world it is the only indication of the drama taking place beneath the placid surface of the water—the miracle of evolution that supports this tender flower.

Matters of life and death

So often to a casual observer the water of pond or river seems devoid of life—save, perhaps, for a sudden hatch of fly flickering from the surface, or the boil of a rising fish. But deep down underneath, among the weed beds, it is teeming with life. There are water mites, beetles, bugs, snails, shrimps and the larvae and pupae of a multitude of flies.

There are also the fish. And it is the behaviour of the water insects that controls the behaviour of the fish that eat them. Often enough, the way in which a fish rises to take an insect tells the

Top
Nymph of mayfly, *Ephemera danica*.

Above
Sub-imago of mayfly, *Ephemera danica*, recently emerged from the empty case below.

nature detective what is happening down in the water, out of sight.

Once flies have hatched and left the surface they influence many other species, too.

In springtime, all along the waterside, a great number of insect-eating birds, both residents and migrants, have claimed their nesting territories and are busy building. Chiffchaff, willow warbler, sedge warbler, reed bunting, wren, wagtail, and many others, including that late arrival from abroad, the grasshopper warbler—whose song is so reminiscent of an angler's whirring fly-reel.

These small birds feed to a large extent on flies that hatch from water: olives, sedges, alders and, for a short time of course on some rivers and lakes, the mayfly—which emerges during the latter part of May or early June.

The mayfly's appearance heralds a period of intense activity. It begins on or about the 26th of May, when mayfly nymphs start to rise to the surface to become winged insects—the first of two stages the mayfly goes through before mating.

Having hatched underwater from an egg into a nymph, and then spending one, two or even three years in the nymphal state, the mayfly emerges from the water and changes into a flying insect known as an imago or 'dun'. This is the insect's first stage as a fly proper. The next day—or, in wet weather, perhaps several days later—the dun sheds its skin and transposes into the final stage: the imago, or 'spinner'.

But now, after this all-so-brief experience of the world above, the mayfly's life is nearly over. Adults live at the most for three or four days. Having reached the spinner stage, mayflies mate and then die—the female, meanwhile, having laid her eggs.

An evening at the waterside during the mayfly 'fortnight' has a magic beyond that of any other insect. The mating flight of adult mayflies is called the 'dance of the spinners' and it is a sight that no observer can fail to recognize. Clouds of male mayflies swirl like snowflakes over the water; and when a female appears, some of them fly after her, often to a considerable height—where pairing occurs.

When mating has taken place, the female lays her eggs on the surface—and dies soon afterwards. As the evening darkens, the water becomes speckled with dead and dying insects, and the trout feed avidly, boiling at the surface as they take fly after fly, gorging themselves at the year's most luscious feast.

Midsummer comes, and the waterside is ablaze with marigold, iris, wild rose, ragged robin and a multitude of other flowers. These are the 'dog' days, the 'lazy' days of high summer. But they are not lazy days for the insect-hunting birds. The wealth of

Ripples spread in rings across the surface of a Hampshire mill pond, showing where a trout has risen to suck in a newly-hatched fly.

summer food coincides with the hatching of young birds and there are many families to be fed.

As the world turns and the chicks grow, the food demands increase. Fully-fledged families keep their parents constantly flying to and from the water with fresh supplies of insects.

Close to a nest almost at water level among the sedge, young coots are being fed on insect larvae and crustaceans. A young coot needs to be fed, but that marvellous swimmer the mallard duckling can feed itself.

Beds of water weeds are trout larders, full of larvae and snails and shrimps. But when swathes of weed are cut in midsummer to drift downstream in rafts, the birds are on to them at once. And other animals, not all of them insect-eaters, take advantage of this unexpected food supply. We may see an elephant hawk-moth caterpillar eating its fill of the lush growth. Or that familiar little riverside character the water vole, which—although often mistakenly called the water rat—is mainly vegetarian.

Fording a river in low water at a place that often shows the surface 'V's of running sea trout. Such low-water runs of fish usually occur at nightfall; but sometimes do not start until the early hours of the morning. To anyone alone at the riverside at night and unfamiliar with this aspect of sea trout behaviour, the sudden violent splashing, as fish force their way through water only an inch or two deep, can be very startling.

Fish that splash in the darkness

The rings and swirls of feeding trout are by no means the only signs that advertise the behaviour of fish. For instance, groups of bubbles indicate the arrival of tench in a baited swim. Salmon and sea trout frequently jump clear of the water. The position of a pike can be pinpointed by the activity of small prey-fish skittering across the surface as they attempt to escape this great predator. Young herrings do the same, when pursued by a shoal of mackerel; and the mackerel, sometimes, when attacked by a school of porpoises. The dorsal and caudal fins of basking sharks are visible as these huge fish cruise steadily along, sieving from the upper layers of the sea the plankton on which their lives depend. In Loch Lomond, powan shoals often come to the surface—which the fish furrow with their dorsal fins. A form of behaviour known locally as 'finning'. It has been said of vendace, too, that they will come to the top like herrings '. . . making a similar noise by their rise and fall to and from the surface'.

But perhaps the strangest and most eerie sound made by fish is the sudden loud splashing of sea trout, as they force their way upstream through shallow water in the darkness.

It is a spate that encourages a daylight run of sea trout and salmon from the sea; but both species show a preference for running at night, continuing to do so even when the river is fairly low. Once a river has dropped below a certain level, however, salmon will not run either by day or night; whereas sea trout can

Sea trout, splashing through the shallows.

A few of the clues that enable us to distinguish between salmon and sea trout.

Salmon (grilse), 6lb. Note the forked tail. Very few spots below lateral line.

Sea trout, 6 lb. Note the square tail. Also, profusion of spots below lateral line.

Tail of 13½lb salmon.

Tail of 11¼lb sea trout.

find their way upriver at night in *very* low water. Some fish, in fact, running in water so shallow that between pools they are often unable to swim and are forced to propel themselves on their sides through what is little more than a trickle in a series of convulsive flops. I have watched them doing this on many a June or July night during spells of dry weather.

On any but the darkest night, the surface bulge can be plainly seen as a running sea trout comes over the sill at the tail of a pool, followed by the long 'V' of its wake as it swims on into the deeper water above.

The sea trout and the otters

Otters seem to have decreased in number during the last decade. Today I seldom see one. But years ago I used to watch otters quite frequently on the little river near my home.

In one particular pool a steep clay bank was used as an otter slide, and early one summer morning I witnessed a very curious incident. It was just after sunrise. The day was calm; the water low, gin-clear and unrippled. I crept upstream along the river bank and from a vantage point behind some broom bushes saw two otters playing in the pool. At least, I assume they were playing. They were running up the bank, tumbling down into the water one after the other and running up again. But what astonished me was the sight of a shoal of sea trout lying seemingly unperturbed not far distant in the pool tail.

Usually, as soon as an otter enters a pool, fish start darting

Murder by the waterside. Who has been done to death? Who was the 'villain'?

An otter 'kill' found on the bank of a Cumbrian salmon and sea trout river. Although little of the fish remains, the shape of head and tail identifies it as a small female salmon.

The skin has been folded back over the tail, untorn, like a concertina.

The bones have been stripped bare. The gills and every morsel of flesh have been eaten out of the head. If we wanted to clean out a fish so thoroughly we should need to use special tools. The otter has done it with only its teeth and forepaws: a wonderful feeding skill.

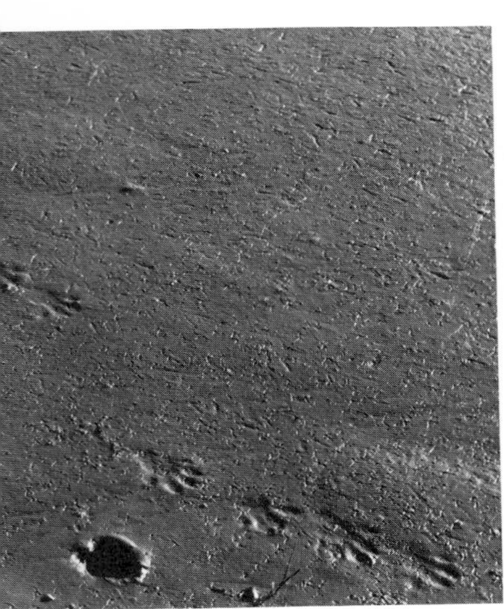

Above right
Another otter meal: freshwater mussels.

Above
Otter tracks on the river estuary near my home.

about in acute alarm. I have seen them do it. But on this occasion the sea trout ignored the otters and continued to lie where they were—utterly different behaviour from that of frightened fish!

To me it was inconceivable that the fish could be unaware of the otters' presence in the pool. Indeed, it seemed, rather, that in some mysterious way they sensed that the otters were not hunting; were simply playing about and therefore posed no immediate threat.

It was uncanny. As though some natural law had been suddenly turned topsy-turvy and was standing on its head. Whether such behaviour is common of the sea trout—one of the shyest of fish—I cannot say. At no other time have I ever watched otters at play in such perfect conditions for seeing fish. Whether such a state of peace would have continued for much longer is also a matter of conjecture, for my Labrador, unable to contain himself, darted forward ... and in a swirl of water the otters disappeared upstream.

Red deer hinds in the snow.

Snow

No animal can conceal its movements when snow is on the ground. In windless weather after a fresh fall there are tracks everywhere, the prints being clearest when a moderate snowfall is touched by frost. Many of the animals are hungry in cold weather and there is usually much feeding activity.

When the weather turns really hard, life can be difficult for many of the birds (during the prolonged freeze-up of 1963, I found a number of redshanks and curlew so weak and emaciated that they were unable to fly. Wading birds are the hardest hit when the ground is frozen). But for a while at least most of the mammals are not too badly off. Hare and rabbits suffer from deep snow, but voles and mice can feed underneath it. Otters can hunt under the ice. Badgers survive winters far more severe than most experienced in Britain—living off their fat, which is specially

Otter tracks (or 'seal') in snow covering a frozen ditch. The rather pointed toes and large unsymmetrical pads make them distinctive and easy to recognize. The dog otter is often markedly larger and heavier than the bitch.

Mallard prints beside a frozen pond.

Mallards born in Britain tend to stay in Britain, but (to take advantage of our milder winters) a considerable immigration of mallards takes place each October from countries around the Baltic and from Russia. Several ringed birds are known to have flown as far as 2,500 miles west.

Migrating mallards can maintain a cruising speed of 50 m.p.h. with no great trouble. Two ringed birds are known to have covered over 500 miles in two days. Another ringed bird travelled 900 miles in five days.

Hare prints in snow.

This magnificent display of wings and tail feathers shows where a raven has landed in deep snow, and walked on towards bottom right.

The raven, biggest of British corvids, seems to epitomize the very spirit of wild places. For many years, ravens have nested on a crag behind my cottage, the distant croaking a feature of quiet summer evenings. Their autumn aerobatics performed in an up-draught on a flank of the crag, often in family formation, are strikingly dramatic.

stored up for hard times. Deer, too, can live off their fat for a limited period—if they start the winter in good condition.

The fox can always make a living. For a few weeks in January, during their mating season, foxes are on the move all the time. Anyone seeing them out in the snow might think they were starving. But not a bit of it. In a prolonged freeze-up they thrive on the smaller creatures that, by now, are dying of hunger. Of course, as soon as their mating season is over, they revert to their normal routine of coming out only at night.

It is after a snowfall that one discovers how many different animals visit unlikely places. The nature detective may be surprised by the number of tracks to be found in town parks and municipal gardens, as well as in the garden at home.

Curious behaviour of a swimming mole

Despite the large amount that has been written about the mole, many aspects of this little animal's behaviour remain a mystery; for instance, no one really knows how it finds it way about. It has indifferent hearing and is nearly blind. Its sense of smell is not too good, either.

But although the mole's pink snout may be poor on smells it is extremely sensitive in other ways, being covered with thousands of tiny raised papillae (known as Eimer's organs), which are richly endowed with nerve endings and, together with the vibrissae—whisker-like hairs that surround the snout—are thought to be capable of detecting minute air currents and changes in air pressure.

If this is so, the hyper-sensitivity of this remarkable snout may go some way towards explaining a curious incident that occurred not long ago in the river near my home.

Heaps of freshly turned earth. Signs that are commonplace to all but the most inexperienced nature detective—or, perhaps, a visitor from Ireland, where the mole is unknown.

Mole hills are formed by the earth that the burrowing animal pushes up out of a newly dug tunnel. Burrowing is not done when the animal is in search of food. A tunnel is simply a blunder-trap into which worms and insects fall, later to be found and eaten by the foraging mole as it patrols its network of tunnels. Digging out a fresh tunnel increases the mole's food 'catchment' area.

The mole lives a solitary life, a male and female meeting very briefly once a year to copulate—and parting immediately afterwards. At all other times of the year a mole will attack and kill, or be killed by, any other mole it meets.

Front foot of a mole—a broad, immensely powerful shovel. It has been estimated that in the process of burrowing, a mole can dig out and push up to the surface about 10 lb of soil in twenty minutes.

Like the mole, the common shrew lives a solitary life, except for mating, and is a good swimmer. It uses grass or leaf-litter runs, or tunnels just under the surface. Its nest is a ball of grass, hidden either below ground or in undergrowth.

On a summer's afternoon, while crossing the bridge pictured on p. 152, I saw a mole tumble into the water from an overhanging bank. On most days I would not have questioned its safety; for as I had witnessed before, moles are splendid swimmers. On this occasion, however, the river was in spate and the mole seemed likely to be swept into a big breaking wave that foamed over the weir just downstream of the bridge.

To my surprise, after a few moments of indeterminate splashing, having been whisked by a swirl of water into mid-stream, the mole turned head-on to the current and started to swim with powerful strokes up-stream away from the weir, its snout pointing up into the air.

It seemed to me that it knew exactly what it was doing. And what puzzled me at the time was how an animal that was virtually blind and unable to get a fix on either bank could possibly tell upstream from downstream—or across-stream, for that matter. In others words, *how did it know the direction of the current*? (A blind-folded human swimmer would be in an exactly similar predicament).

But although, gradually, the current became too strong for the mole, and it began to be carried backwards towards the waterfall, it continued to point an upstream course.

Extraordinary.

Determined to find out whether what had happened was merely a coincidence, I ran downstream, waded into the river and, when the mole arrived at the lip of the weir—still paddling hard upstream, but travelling tail-first downstream—I rescued it.

Then I took it back to the bridge—and dropped it in again.

Once more the mole positioned itself facing upstream stemming the current for a time until, eventually, as before, it began very slowly to be carried downstream, tail-first, towards the weir.

Again I rescued it. This time, however, feeling it had done enough, I offered it freedom on a patch of bare earth nearby.

Instantly it started to delve, working its spade-like forepaws with a breast-stroke action at astonishing speed. I had thought the little creature to be exhausted . . . but in a few seconds it had disappeared underground.

Thinking about it later, having discovered more about the mole's anatomy, a possible explanation occurred to me. If the sense organs of a mole's snout are delicate enough to detect the tiny draught on its face while being swept downstream by a four or five knot current, they might, perhaps, enable the animal to position itself in *relation* to that current.

If the mole can really achieve this, it can do something that in all probability no other animal can do. Certainly, no *human* swimmer could detect a current, however strong—let alone sense its direction—unless there was some stationary object in view by which the swimmer's 'drift' could be assessed.

Homework

Once we can identify certain species of animals and can recognize some of the signs they leave, we can visit suitable new locations and, by tracking, discover which animals are using those grounds and what they are doing there. All of which is a step towards observing the animals themselves. But if we are to watch the natural relaxed day-to-day behaviour of wild animals we must cultivate stealth and fieldcraft, for it is always the *unseen* observer who sees the most.

Wild creatures go about their daily chores, unafraid, only when unaware of our presence. While we are walking openly in the countryside, animals can all too easily see or hear or scent our approach. We ourselves observe very little—except fear!

From the far side of a tree, woodpigeons swoop away from us with a clap of wings. Quacking in alarm, mallards soar from the reedbeds and disappear across the marsh. Rabbits hop into their burrows with warning thumps. Mice rustle under the dry leaves. An adder glides unseen beneath shadowy brambles. Squirrels vanish before our very eyes, putting branches between themselves and us. Even an animal as large as a red deer allows us little more than a glimpse of its white 'target' as it melts away among the trees.

Apart from noting the various alarm reactions, there is not much of interest in any of this. We have made contact only with frightened animals, and many of those have fled unseen and unsuspected.

Successful observation of wild animals depends largely on our ability to conceal ourselves; either by using natural cover such as bracken, long grass, reeds, bushes, banks, hedges, rocks, ditches or trees—to stand motionless beside a tree trunk works well, until we begin to fidget—or by the use of hides. These can be forest 'high seats', often used for watching and filming deer; screen hides made of wire netting thatched with natural vegetation, reeds or rushes, which are often used on wildfowl reserves; or portable camouflaged canvas hides that can be erected quickly in any chosen spot.

Previous page
View from my kitchen doorway. Cumulus clouds cover the top of Scawfell in this picture taken at midday. But sometimes at daybreak the scene is breathtaking: the sky behind the mountains streaked with red and orange flame, the valley a ribbon of mist tinted by the rising sun. With the air full of bird song and the wild crying of curlews coming from a distant fellside, the centuries seem to peel away and one may imagine a party of ancient Britons clattering down towards the river from their settlement high up behind the Latterbarrow Crag.

A roe buck's fraying-stock. The bare trunk (top arrow) shows where the buck has rubbed off the bark with his forehead and antlers. While fraying, the buck has pawed at the ground with a forefoot, leaving traces of scattered moss and soil (bottom arrow). Scent from glands in the buck's forehead has been left on the tree: a signal to other bucks that: 'This territory is occupied!'

Roe bucks frequently use river banks, rides and paths through woodland as part of their territorial boundaries. Bushes and young trees used as fraying-stocks are scattered all round a buck's boundary, and he will visit them regularly—often morning and evening—to renew the scent that claims them as his own.

Once these signs have been found, an observer who has the patience and fieldcraft to wait and watch unseen (and unscented) by the deer, can identify that particular animal by its antlers.

Unlike red deer, roe tend to live in small family groups, each family remaining in one particular area. A family may consist of the male (buck); the female (doe), plus the current year's young (fawns). Does often rear twins, so that the family group may hold as many as four animals.

A mature red stag, in velvet, and a two-year-old staggie with only the beginnings of antler-growth.

A motor car can be a useful hide when parked at a strategic place by some lonely roadside. Most animals pay little or no attention to vehicles that are stationary, or moving very slowly.

But there is another, very simple, form of concealment that is available to almost everybody. In other than totally built-up areas, potentially good observation hides are the houses we live in. A surprising variety of wild animals visit domestic gardens, many of which are by no means in the depths of the country, and much interesting behaviour has been recorded from them. Needless to say, however, those of us who live or spend some of our time in fairly isolated homes, have even better opportunities, and my own home is no exception.

A beautiful example of camouflage. Like this young roe, most deer show camouflage spots during the vulnerable first few weeks of their lives.

Remains of a roe buck that has died a natural death. Its bones have been stripped clean by foxes—with some assistance, doubtless, from buzzards, crows and ravens. Much information about the animal can be gained from examination of its teeth and antlers.

Roe deer rut between the third week of July and the end of August. Signs of the rutting stands can often be found. When the bucks are driving does, circles tend to be formed in the ground where the earth is heavily cut up by the trampling hooves.

The river valley I live in, like many similar valleys, has a narrow strip of cultivated ground along the valley bottom stretching up into a wilderness of fells. It has a rich variety of wildlife, and few of the animal species that inhabit or visit it cannot be watched from my cottage window or from vantage points nearby. Like any other wild or semi-wild place (estuaries in particular) a valley such

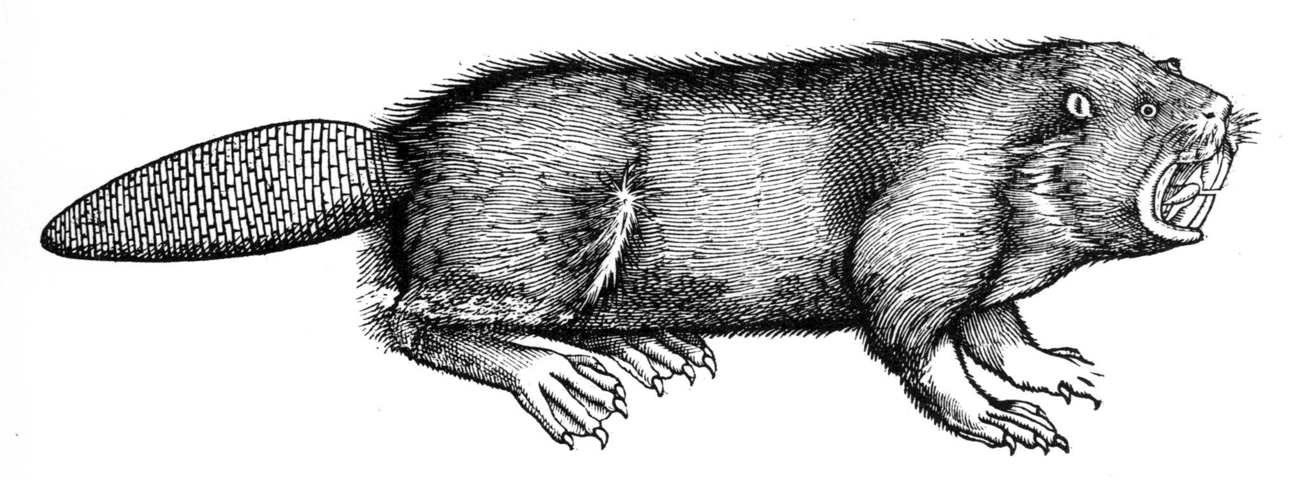

The beaver, greatly prized for its skin by the medieval nobility, seems to have survived in Wales and Scotland until the 14th and 15th centuries, although in England it was probably extinct before Anglo-Saxon times. This comparatively accurate illustration comes from a dictionary of animals published in 1607.

A roe buck's territorial boundary stretches along the near side of this pond. At top left is Latterbarrow Crag behind which was the ancient Briton settlement of Barnscar. In those days the valley bottom would have been swamp and scrub, the slopes beyond densely wooded—holding, among other species now banished from Britain, bear, wolf, boar, lynx and beaver.

Roe deer numbers fell drastically during the deforestation of Britain and the species was reintroduced in the south of England two hundred years ago. In the Border counties, however, roe never completely died out. Since man has existed, roe deer have probably lived in the countryside pictured above.

Roe buck, fraying.

Sapling pushed back by fraying roe buck.

Roe buck. The drawing shows the position of the animal's scent glands, used for territorial marking. Also shown is the 'target': a white patch on the rump used as a recognition signal for the young, when the parent is alarmed and retreating from danger. It warns the young deer to 'Follow me!'

as this is a wildlife 'hotel'. It houses a few residents and a host of visitors that come and go according to the seasons. And whatever the weather or time of year it is always beautiful.

As I write, at daybreak, the sky is catching fire behind Scawfell. The river is a ribbon of mist glowing pink in the early summer sunlight. Larks are singing, hidden somewhere in a pale blue sky. The bubbling cries of nesting curlews come distantly from the fell.

A magpie hops across the tiny lawn outside my window, stops

to look and listen, then hops on again. A jay screeches from the hedge. Behind the chorus of smaller birds is a background of cawing rooks busy in nearby treetops.

In the moss, down towards the river, a roe buck emerges from clumps of bog myrtle, visible at first only by its white target as it moves slowly along nibbling the tips of bramble bushes that fringe the moss.

Later, when I take my Labradors along the river bank for their morning exercise, I shall see the signs that proclaim this particular buck's territorial boundary: saplings, or bushes, with the stems peeled clean of bark, which he has rubbed with head and antlers while leaving scent from his forehead glands as a signal to other bucks: a warning that tells them 'This is *my* pitch. Keep out!'

At the base of each 'fraying-stock' are other signs that help to tell the story: heaps of earth and grass and moss, scuffled out by the buck as he pawed the ground while fraying.

In another month or two, at rutting time, the challenging barks of rival bucks will echo through the darkness.

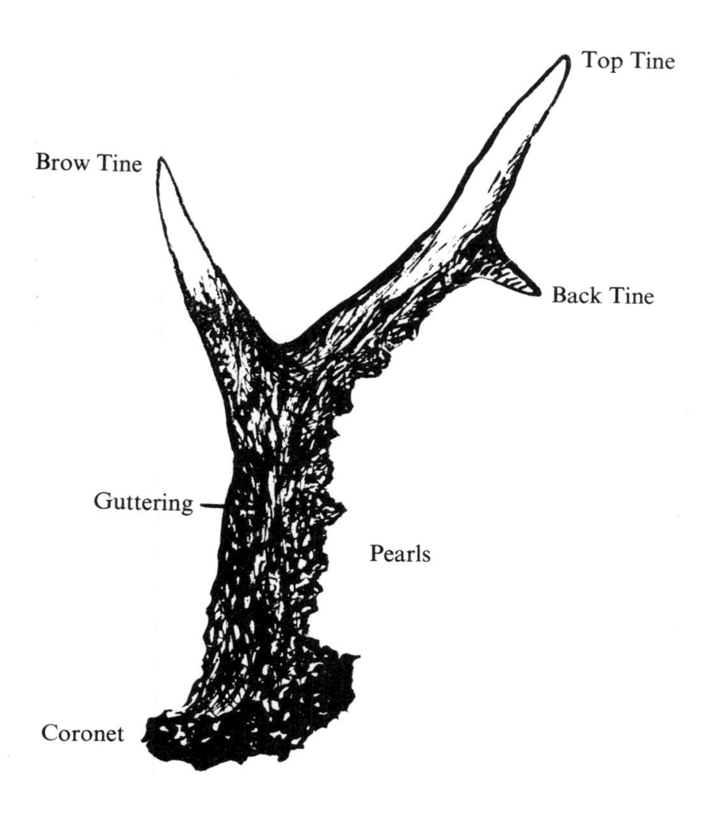

Roe buck antlers and technical terms. Like other deer species, only the male carries antlers.

Information from antlers

Above left
Antlers of a young capital roe buck, four years of age. The head shows excellent symmetry; first class pearling and good thick coronets. As with a stag, a buck can be recognized each year by the shape and size of his antlers.

It is fascinating to reflect that fossilized antlers similar to these have been found in Britain together with the remains of hippopotamus, rhinoceros, wild horse, hyena and bear; and although they are 400,000 years old there is hardly any difference between the roe of pre-historic times and the roe deer of today.

Above right
Roe buck antlers in 'velvet'. Like other male deer, a roe buck casts his antlers each year and grows a fresh set, the process taking three or four months to complete. While the new antlers are growing they are covered with a sheath of skin and hair and blood-vessels. Until the antlers are fully grown this 'velvet' is very sensitive, but when the antlers have hardened the sheath of skin dries up and seems to become irritating. Like a stag, a buck then rubs his antlers clean against young trees—which are almost invariably killed in the process.

Growing new antlers puts a strain on the animal's mineral reserves. To gain calcium, deer often eat their cast antlers—which is the reason why so few cast antlers are found lying about.

Opposite top left
Distorted antlers. Damage to antlers is caused either by injury or by some form of internal disease. Frightened deer—chased, perhaps, by stray dogs—frequently crash into fences or other obstacles. Antlers that receive a hard blow when soft and in 'velvet' are usually broken or split. If the pedicle, out of which the antler grows, is seriously damaged, as often happens, that particular animal will always produce a deformed antler in subsequent years.

Such malformations are easily recognized and are used by the woodsman not only to identify an animal, but to decide which bucks should be culled when he is reducing numbers to the accepted level that his forest can maintain.

The curled-up antlers of an old buck suffering from lungworm. Marked down to be culled, he was easily recognized by the ranger.

Above right
Antlers of an old diseased buck.

The saplings (arrowed), which have been used as fraying stocks, were planted at the same time as the trees seen in the background. Damage to the bark has inhibited their growth.

A roe deer 'slide' on the side of a clay bank. Faint parallel lines show where the animal has slithered down on its toes.

Roe deer print with (below and going in the same direction, i.e. right to left) otter tracks.

Roe deer again? No! Sheep.

Grizedale Forest

Like the rest of Britain, the bracken-covered fellsides of the English Lake District were once dense forest and rich in wildlife: the haunt of lynx, wolves and boars, kites, pine martens, golden eagles and a host of other birds and mammals. The Chronicle of Lanercost describes the Forest of Cumberland as 'a goodly great forest, full of woods, red deer and fallow, wild swine and all manner of wild beasts'—which is typical of the Royal Forests of the time, when sport was the main consideration.

But that was long ago. Already in the middle ages, forest clearance had begun, chiefly to promote grazing land for sheep. By Elizabethan times many of the fells were partially denuded of trees, and during the next two centuries the Iron Masters finished the job when they cut down the remaining trees for charcoal.

As the forests dwindled the wildlife dwindled too. The questing muzzles of sheep prevented the regeneration of trees and gradually the bracken took over with a host of weeds.

Sphagnum bogland with tarn in Grizedale Forest—haunt of deer and wildfowl.

Opposite
The jay—known as 'the forester's friend'.

When acorns loosen in their cups, oak trees provide a harvest for many wild creatures. Pigeons, pheasants, ducks, squirrels, mice, badgers, deer and pigs feast on acorns in the autumn—the jay in particular. Some of the jay's acorns are eaten on the spot; but many are pouched, carried away and hidden in the ground to form a winter store-house. Like that of fox and squirrel, the jay's habit of hoarding seems to be a response to a super-abundance of food.

Of course, jays do not recover everything they bury. Many acorns are left in the ground—and a few germinate. Which is how some of our oak trees get planted out.

The wild boar, acorn-eater of the old Grizedale oak forests, was extinct in southern England by the time of Charles I, but may have lingered in northern Britain until the end of the 17th century. A re-introduction was attempted *c.* 1702 when General Howe imported a number of boars from Germany, and released them in Wolmer Forest. According to Gilbert White, 'The country rose upon them and destroyed them.'

The wolf lasted longer, roaming England until the reign of Henry VII and Scotland until as late as the 18th century.

According to F. Fraser Darling and J. Morton Boyd in *The Highlands and Islands*, 1964: 'The story of the wolf (*Canis lupus*) in the Highlands is important because the animal was responsible for a good deal of the later history of the destruction of the forests. Clearance of the forest by burning was doubtless the easiest way of restricting the wolf's range. The last wolf of Scotland is said to have been killed by one, Macqueen, on the lands of Mackintosh of Mackintosh, Inverness-shire, in 1743 . . . we are dubious of all stories attributing man-hunting to the wolf, although wolves were plentiful and hungry enough to cause people in the Highland areas to bury their dead on islands off-shore. Examples of such islands for which this tradition exists are Handa, Sutherland; Tanera, N.W. Ross; Inishait, Loch Awe; and Eilean Munde, Loch Leven, Argyll.'

Opposite top
Tracking with John Cubby, Head Ranger of Grizedale Forest. Many people (including myself) would have missed this evidence of red deer feeding—shown in close-up in the picture below. But not Cubby, whose eyes from long experience are trained to notice the slightest signs. The gap between amateur and professional is just as marked in tracking as in any other profession.

Opposite bottom
Clump of grass eaten by red deer. A very faint sign indeed, but obvious enough when pointed out by an expert.
Note: Only the tips of the grass stems have been nibbled. Had the grass been eaten by cattle, it would have been grazed much shorter.

Walking too fast and trying to scan too wide a field is a common mistake made by many people who go into the countryside to seek signs of wildlife. It is so easy to tread on clues that lie (literally) underfoot: crushed or cropped vegetation, feathers, footprints, scrapes, droppings.

It was not until the early 18th century that new plantations began to be formed with an ever-increasing introduction of coniferous species—which eventually prompted Wordsworth to liken the local larch woods to 'a sort of abominable vegetable manufactory!'

And some people might think the same about many of today's Forestry Commission plantations. But there are exceptions. And one notable exception is the Forest of Grizedale, whose 7,000 acres stretch between Lake Windermere and Coniston Water.

Red stag dropping typical of rutting time (October/November).

Sheep droppings.

Red hind droppings.

Above
Faint signs of flattened grass at the base of a tree (arrowed) show where a roe deer has been couching.

Above right
A larger and more obvious area of flattened grass: the couching place of a red deer.

Right
A glade in a new plantation greatly favoured by red and roe deer.

Opposite
Short-tailed vole. Most of the forest inhabitants leave their signatures on trees. At various heights, bark will be peeled off the trunks by deer or sheep or rabbits. In the winter, considerable nibbling damage is caused by the vole—a most destructive little animal. During late spring and summer it can eat its own weight of fresh grass daily. When grass is lacking, it can kill a young tree by gnawing round the base. In turn, it is eaten by many predators: chiefly owls, kestrels, buzzards, weasels, stoats and foxes.

Fresh fox dropping on hummock. The use of little hillocks and mounds as lavatories is typical of fox behaviour.

Opposite
A wet wallow used by red stags at rutting time. The arrow indicates a patch of ground where stags have rolled about after wallowing.

Red deer rutting stands have been known for generations. These areas are usually in the vicinity of a wallow, and the ground is very heavily trampled. Rutting takes place mainly during the month of October. On a still night, the roaring of the stags echoes through the forest and lends dramatic emphasis to the autumnal scene.

'Grizedale' means, literally, 'The valley of swine'. It was named by the Norsemen who settled there in the 10th century, when it was virgin oak forest. Then, wild boars fattened on the acorns and wolves kept the number of deer in check. Today there are neither wolves nor boars, but the name has lasted, and so have the red deer.

It is a place of great interest—where I have spent many happy hours observing and tracking—and has featured in several of my radio and television programmes. It holds some of the best remaining oak woods in the Lake District, with an excellent herd of red deer and many roe. Since it is close to my home, and therefore a forest I know quite well, Grizedale has been used to illustrate these few notes on deer-tracking.

Riddle of the red deer

Two signs of red deer activity that are of considerable interest, both to amateur and professional observers alike, are the wet and dry wallows.

The wet wallow is a small area of mud and water in which a stag will immerse himself and roll about, rather like a pig. This curious behaviour takes place just prior to and during the rut, and nobody knows the reason for it.

One theory is that the wallow acts as a 'mud bath', cleansing the stag's skin of parasites. Another theory suggests that the mud, having dried and flaked off, retains the stag's scent—thus helping him to mark out his territory and warn off other stags.

Perhaps the most likely reason is the one offered by Bill Grant, Chief Forester at Grizedale, in my programme: *Wildlife and the Forest*.

'Stags, during the rut, seldom eat and are in a constant state of excitement. There is much roaring, and incessant movement as they continuously round-up their hinds and ward off would-be territorial invaders. We may imagine that the effect of wallowing in soft, cold mud must do much to cool the fire of their ardour!'

Close-up of arrowed portion of wet wallow showing vegetation flattened and matted with mud by rolling stags.

Just a bare patch of earth in a forest clearing. It seems quite ordinary. Something that most of us would walk past without a second glance. And yet, like the little mud pool already described, what a fascinating and baffling story it tells.

John Cubby examines a dry wallow just prior to the red deer rut.

A dry wallow is a patch of dry earth, usually on the side of a slope, on which hinds roll about. Again, very curious behaviour. It has been suggested that the hinds do it to rid themselves of ticks, which can be just as troublesome to deer as to sheep and other animals. Whether this is really the answer, nobody knows. Like that of the wet wallow, the function of the dry wallow—if it has one—remains a mystery.

Opposite
The long bare patch on the side of this tree shows where red deer have been peeling off the bark. Deer have incisors only in the lower jaw, so when peeling bark they use these bottom incisors and strip *upwards*. The teeth marks can be plainly seen from close examination, and these prevent confusion between bark-stripping by deer and the slashes made by a forester when he marks a tree for felling.

Most animals are quite harmless to tree growth, but the forest dweller that made its mark in such dramatic fashion on the trunk of the tree opposite may have signed the tree's death warrant.

Above
Red deer will strip bark at any height within reach. Here the base of a tree has been peeled. Other animals such as sheep, voles and rabbits also take bark low down on a tree. In this case, however, teeth marks in the wood prove that a red deer was responsible.

Right
Red deer peeling scars from which resin has started to flow. Deer dislike the taste of resin and will not continue peeling bark once the flow starts. A recent Continental method of preventing deer from stripping bark is to prick the tree and start a small flow of resin—sufficient to repel deer without causing ill effects to the tree.

The red stag's curious habit of wallowing in muddy holes was noted centuries ago—and sometimes put to grisly use. Craven, in *Recreations in Shooting* (1846), quotes an early nineteenth century highland stalker:

'We passed during the day several forest-baths in full use—that is, moss holes, where the stags plunge up to the neck, and roll about to cool themselves, in summer and autumn. When they come out again, black as pitch, they look like the evil genii of the mountain. In former times, poachers used to fasten spears with the points upwards in these places, and when the stag threw himself into the hole, he became impaled.'

Lord Lovat, in *Shooting* (1887), also makes the point: '. . . in these early days stalking was little practised. Driving, coursing with men and dogs, mobbing deer among the rocks and killing them in the lochs, were the blunt methods employed. Even pitfalls and snares, and an arrangement of sharp spikes in the bathing pools, were often used . . .'

Death of a tree.

Like the broken mast of an old-time sailing ship, a pine tree measures its length beside four feet of shattered stump. What has caused this accident in the forest? Lightning? A gale force wind?

No. The destruction of this tree is due not to the elements but to the appetite of a red deer. Bark peeling similar to that shown on p. 138 has allowed fungus to enter the trunk and caused it to collapse. The upper part of the trunk (arrowed) has snapped off and lies where it fell, pointing away from the camera.

Thought to have been introduced to Britain by the invading Romans, the fallow deer (male, *buck*; female, *doe*; young, *fawn*), is the commonest park deer and feral in many districts—especially the New Forest. Once a favourite of the royal hunt, it was protected by William the Conqueror's Forest Law together with the indigenous red deer, roe deer and the wild boar.

Signs of damage

When trees are planted and areas of woodland created, many species of wildlife are attracted into those areas. By sign-reading and observation we can identify these animals and discover which of them are harmful to tree growth and which are harmless, or even helpful.

Certain species can do great damage—the oak leaf roller moth

More signs of feeding damage. This time by roe deer. Roe are very partial to the branches and leaders of young trees.

Unlike red deer, roe deer are territorial and live in small family groups, keeping very much to their own part of the forest. In consequence of this, they are constantly browsing in the same areas of forest and upon the same trees. They feed mainly at night and have regular feeding patterns. Choosy feeders, they like a varied diet and tend to frequent glades where there are plenty of mosses and sedges and lush grass in the damp spots, with young trees round about to browse on. The severity of their browsing (arrowed) is an indication of the density of the local roe deer population.

can defoliate an oak completely—whereas insect-eating birds, such as the pied flycatcher for example, are very helpful. Again, since a young plantation can be destroyed by the short-tailed vole, the controlling value of predators such as fox, stoat, weasel, buzzard, tawny owl, kestrel and other species that prey on the vole, can be readily understood.

Deer can cause considerable damage both to trees and to agricultural crops. But (apart from the fox which will very occasionally kill new born fawns) deer have no natural predator left in Britain. If we wish to grow crops and trees and at the same time live in harmony with deer, some control of deer numbers is obligatory.

The key to effective deer control is the forester's ability to read tracks and signs, from which he can assess the number of deer in his forest, their species and sex, the areas they frequent and their behaviour within those areas. Then, from observation and careful selection he can decide which animals to cull.

Typical roe deer feeding damage to young Norway spruce. A distinct 'browse line' can be seen (arrowed), the point above which a roe deer cannot reach (a fully grown buck is only about twenty-six inches high at the shoulder). Below the browse line the branches have been nibbled away.

Above
Spruce leader nibbled by deer.

Right
The double leader is typical of deer damage, occurring when the original leader has been nibbled off (see picture above).

A roe deer feeding ground. Plenty of young trees to browse on, with sedges and mosses in the damp spots and nearby cover to bound into if disturbed.

Judicious culling achieves a dual purpose: it protects tree growth and agricultural crops and ensures a healthy stock of deer—for it replaces natural culling of the weaker animals by wolves, which were finally exterminated in Britain little more than two hundred years ago.

To implement this policy, modern forestry employs rangers who are expert sign-readers. The system of deer control in Grizedale Forest is an example of nature detective work at the highest professional level.

Fencing

In theory, forest enclosure may seem a simple enough matter; but in practice, with many miles of fencing to erect and maintain, it is a complicated and expensive business. Deer are phenomenal jumpers and fencing to less than six feet high is not effective.

Roe deer in particular are very persistent. Sometimes, alongside a successful fence surrounding a new plantation, roe that have been trying to get in time after time will leave a beaten pathway, just like a sheep trod. The slightest weakness in a fence, and deer will quickly break their way through.

Roe deer, being territorial and usually much more numerous than red deer, do more damage to the woodland. Red deer, not being territorial, are generally on the move and their browsing areas are more scattered.

The habits of roe are very regular. A buck usually controls a territory of perhaps twenty to twenty-five acres and, if not disturbed, remains there indefinitely. Roe have a daily feeding pattern, coming down off the tops of the forest hills in the evening to feed on the lower, lusher slopes. During the first part of the morning they return to the comparative safety of the upper slopes to lie up or browse. It is said that you can almost set your watch by their daily movements along the same regular routes.

The months of April and May are probably the best for watching roe. At this time of the year they are eager to feed on new growth—especially after a hard winter. If undisturbed they will continued to feed later in the morning and seem not to be quite so shy.

Young tree that has been used as a red stag's thrashing stock, and destroyed in the process. A stag seems to derive great pleasure from belting young trees. He will spar with a sapling just as a boxer spars with a punch-ball—beating it down and letting it spring back again, time after time.

A forest punchball

Of all times for seeking signs of deer and for observing the animals themselves, the early morning is the best. But however good the light, deer can be very difficult to distinguish from a background of trees and undergrowth. A roe buck melts into the bushes. A red hind crouches in the grass unseen until suddenly, she winds the observer and ghosts away like a shadow among the trees.

Even in the slanting light soon after sunrise, which shows up footprints to their best advantage, tracks are not easily noticed during a spell of dry weather when the ground is hard.

Irrespective of weather conditions or intensity of light, some clues are always very faint. Footprints in damp grass will be just discernible—until the dew has dried. Other clues are missed simply because we fail to recognize them for what they are—indeed, except by a practised observer who understands the significance of what he is looking at, some of the most obvious examples of deer activity may be ignored.

For instance, the red stag's thrashing-stock—a young tree against which the stag has cleaned the velvet from his antlers.

Note. Some red stags fail to grow antlers. These animals are known as 'hummels' and are usually bigger than normal stags. Their extra size is probably due to certain minerals being used in body building instead of antler production.

The hunting of the deer, from Craven's *Recreations in Shooting*:

'The golden era of deer-hunting, in all its branches, must be placed at that epoch of an early history, when every hill and mountain glen, every forest pass and sylvan flat, every bog and morass was tenanted by its *ferae naturae*. The wild deer then held state in the chase as the noblest of animals . . . It is easy for the mind to picture the British Isles in the days of yore, as an aggregate of forest and moorland, thinly bescattered by the habitations of man . . . The general hunting matches of the Norman era prove the multitudes of deer ranging at will in the royal chaces; a thousand have been killed at a single match . . . Already at the Norman conquest the red deer had been hunted to the thinning of their species, and our invaders did their utmost for its preservation and increase. They exacted heavy fines from those who trespassed on deer-enclosures, and life for life, if premeditated, was the doom awarded to the [unauthorized] biped who slew the quadruped.'

Regarding the taming of deer, Craven writes: 'It is said that the red deer may be usefully domesticated, although with more difficulty than the other species. Martial relates of a deer, that he was used to the bridle; and Montaigne alludes to the famous

Opposite
This young larch has been used as a thrashing stock and completely uprooted. The trunk below the ring of bark (arrowed) has been frayed clean by the stag's antlers. Above the ring, where normally it would have been out of reach, the bark has been stripped off by the stag's teeth and eaten. The fresh appearance of the fraying indicates that the incident occurred only a short time before the photograph was taken.

Above
Broken branches and stripped bark point the trail of a red stag. He has stamped along thrashing off his 'velvet' against every young tree he passed.

present made to the Emperor Maximilian, of a deer swifter than a barb, that bore both saddle and bridle!'

Craven also mentions a tame stag 'kept at a shooting-lodge of Lord Breadalbane's, which attacked all who came near it, except the foresters . . . no one dared to pass his haunt unless he knew them.'

It is largely from sign reading that the woodsman builds up a picture of deer activity in his forest. When deer move to and from their feeding grounds they use regular routes. The photograph shows a deer 'run' leading down from a higher part of the forest. From droppings found on the trail we could tell that it had been used recently by a red stag. The presence of flies on the droppings proved that they were very fresh and that the stag must have passed this way only a short time before.

The deer trail crosses a gutter beside a forest road, and over the road itself . . .

. . . alongside twenty-year-old holly bushes that have been eaten down almost to knee height . . .

. . . across the bed of a tiny stream . . .

The Earl of Marr's deer hunt in the year 1618, from Taylor's *Pennilesse Pilgrimage*:

'The manner of the hunting is this: five or six hundred men doe rise early in the morning, and they doe disperse themselves various ways, and seven, eight, or even ten miles compass they doe bring or chase in the deer in many heards (two, three, or four hundred in a heard) to such a place, as the noblemen shall appoint them; then, when the day is come, the lords and gentlemen of their companies doe ride or go to the said places, sometimes wading up to the middles through bournes and rivers; and then, they being come to the place, doe lye down on the ground till those foresaid scouts, which are called the tinckell, doe bring down the deer; but as the proverb says of a bad cook, so these tinckell-men doe lick their own fingers; for besides their bows and arrows which they carry with them, wee can hear now and then a harquebusse going off, which they doe seldom discharge in vain; then after we had stayed three houres, or thereabouts, we might perceive the deer appear on the hills round about us (their heads making a show like a wood), which being followed close by the tinckell, are chased down into the valley where wee lay; then all the valley on each side being waylaid with a hundred couple of strong Irish greyhounds, they are let loose as occasion serves upon the heard of deere, that with the dogs, gunnes, arrowes, durks, and daggers, in the space of two houres four-score fat deere were slaine, which after were disposed, some one way and some another, twenty or thirty miles; and more than enough left for us to make merry withal at our rendezvous. Being come to our lodgings there was such baking, boyling, roasting, and stewing, as if cook ruffian had been there to have scaled the devill in his feathers—the kitchen being alwayes on the side of a banke, many kettles and pots boyling, and many spits turning and winding, with great varietye of cheere, as venison baked, sodden, roast, and stu'de; beef, mutton, goates, kid, hares, fish, salmon, pigeons, hens, capons, chickens, partridge, moorcoots, heathcocks, caperkillies, and termagents, good ale, sacke white and claret, tente (or alligant), and most potent *aqua vitae*. All this, and more than these, we had continually in superfluous abundance, caught by faulconers, fowlers, fishers, and brought by my Lord Marr's tenants and purveyors to vitual our camp, which consisted of fourteen or fifteen hundred men and horses.'

. . . through open woodland, past another holly bush . . .

. . . that has been nibbled almost to the ground. Every deer that comes this way pauses to have a bite to eat, before moving on . . .

. . . down towards the lush feeding grounds of distant pastures . . .

. . . that lie on the other side of this sheep and rabbit fence. This spot was once a jumping point; but, although they are wonderful jumpers, deer prefer to go *through* a fence whenever they can. This hole has been made very recently.

Jumping points are commonly where the forest adjoins a plantation of young trees on which the deer are browsing. As examination of the slot marks in muddy ground will quickly prove, a jumping point is often used by more than one species of deer. In Grizedale Forest, red and roe deer.

'Rough-footed Scots'

One of the uses of the red deer, besides food, was once explained by a Scotsman to King Henry VIII.

'We go a-hunting, and after that we have slain a red deer, we flay off the skin bye and bye, and setting of our barefoot on the inside thereof, for want of cunning shoemakers, we play the cobbler, compassing and measuring so much thereof as shall reach up to our ancles, pricking the upper part thereof with holes, that the water may repass where it enters, and stretching it up with a strong thong of the same above our ancles. So, and please your Grace, we make our shoes. Therefore we using such manner of shoes, the rough hairy side outwards, in your Grace's dominions of England we be called "Rough-footed Scots".'

The evening brings its own feeling of solitude, but one is never really alone. At dusk, as the daylight animals are going to bed, the hunters of the night are stirring, and although few of these nocturnal animals will be seen, most of them will be heard.

From our cottage in the Cumbrian fells my wife and I set out at dusk to fish for sea trout, fording the little rain-starved river at a place where roe deer, too, cross over, or pause to drink. No signs of deer are visible on the shingle (except for occasional droppings) but slot marks in bankside mud lead to the spot. And here, roe can sometimes be seen at nightfall. The low-lying meadows beyond are the hunting grounds of a local barn owl. And after dark the tremulous hooting of tawny owls comes wavering across the valley from the wooded fellside—an eerie 'background music' to sea trout night fly-fishing.

'Goatsucker'

'The nightjar holds in its scientific generic name, *Caprimulgus*, the slander of the ancient Greek *aigothêlas*, 'goatsucker', for the bird which pastoral peoples saw in the dusk catching beetles and other insects attracted by the rank smell of the goats. Still known locally in England by that name of 'goatsucker', in Spain as *chotacabras*, in Germany as *Ziegenmelker*, and in other countries by names of equivalent meaning, the beneficent 'fernowl' exemplifies the superstitions connected with so many of the birds—woodpeckers, kingfishers, cuckoos, and others, as well as nightjars—once grouped together as the Order *Picariae*, which in appearance and habits depart from the standards of most birds of similar size. The swift, for instance, is named by the countryman as 'devil-bird', 'deviling', or plain 'devil'.

'It would have been as difficult to convince the old-fashioned game-keeper that the kestrel and the barn-owl are far more useful than harmful to man's interests, as it would have been to persuade the ancient Arcadian or the 18th-century English farmer that the nightjar, with its bill so wonderfully adapted for hawking for insects, was quite incapable of milking his beasts.

'If Gilbert White did not explicitly attack the absurdity of the name *goatsucker*, as a good naturalist he dissected the crop of a nightjar and found it "stuffed hard with large *phalenae*, moths of several sorts, and their eggs, which had no doubt been forced out of those insects by the action of swallowing". That was in, or prior to, 1776. Twenty-eight years later Bewick, in his *History of British Birds*, named the bird *night-jar* and spoke of it as "a great destroyer of the cockchafer or dor-beetle, from which circumstance in some places it is called the dor-hawk'. Almost simultaneously Montagu was still writing the libellous name of *goatsucker*, noting *nightjar* as a provincial name. But Bewick's influence prevailed, and from the time of Yarrell's *History of British Birds* (1843) it is "goatsucker" that has been relegated to the category of provincialisms.'

Stephen Potter and Laurens Sargent, *Pedigree: Words from Nature*.

The nightjar, otherwise known as dor-hawk, fern owl and goat-sucker.

Things that go bump in the night

It is not of course only from their *visual* tracks and signs that animals can be identified. We can detect and name many unseen species simply by the use of our ears and, sometimes, our noses—although the human nose is a feeble organ compared with that of most animals.

Birds are an exception. On the whole they seem to rely very little on sense of smell. But many mammals, fish, insects and

Tawny owl

reptiles have scenting powers that to us seem almost miraculous. Some species can *navigate* by their noses—like the salmon, which is guided from coastal waters to its parent river by the particular odour of that river. Others, the roe deer for instance, communicate by smell, using their scent glands for marking territorial boundaries. And some use scent for attracting a mate: the female Emperor moth's smell can be detected by a male half-a-mile away. And some species hunt mainly by smell. When I watch my dogs nosing into thick cover as they pick up the scent of hidden game, I marvel at the acuteness of their smelling sense.

Descendants of the wolf, they are natural hunters and in the wild state must have depended largely on their noses for getting a living.

I suppose that at some time in the distant past our own sense of smell must have been much more acute than it is today.

'They haven't got no noses,
The fallen sons of Eve . . .'

wrote Chesterton. And he was right. Nevertheless, there are animal species with odours so strong that even *we* can smell them. The fox, for instance; an angry mink; a polecat; a weasel. And who can fail to wind (or will ever forget) the pungent stench of an undoctored tom cat.

Even so, it is mainly from the sounds they make that we can identify most unseen animals. Bird song, for example, is strikingly individual. How different are the calls of partridge, cuckoo, jackdaw, jay, grasshopper warbler, pheasant, yaffle, lapwing, robin, wren—to name just a few of the many daylight species. There are exceptions, of course. The delicate 'tuee, tuee' of the willow warbler is very similar to the call of the chiffchaff. The rook's 'caw' can be mistaken for the crow's. The 'yelping' of the various grey geese can be confusing to any but an expert ear. The starling's gift of mimicry can be misleading. But by far the majority of calls are unmistakable. Who, for instance, can confuse with any other species the wild mewing of courting buzzards; a raven's croak; the mournful piping of redshanks as they flicker across the saltings; or (for me perhaps the most evocative sound on earth) the curlew's haunting cry, echoing from distant fells at summer daybreak?

But the animals that go about their business during the hours of daylight are seldom heard except between dawn and dusk. The darkness has creatures of its own—and its own 'soundtrack'.

When daytime animals have gone to bed, the night prowlers emerge and introduce an entirely different set of sounds. And to an unfamiliar and apprehensive ear some of these night calls can be alarming. Indeed, any would-be nature detectives of timid disposition, who propose to walk alone at night, would be wise before setting out to acquaint themselves with some of these nocturnal noises.

The person who goes into the countryside at night knowing nothing of the creatures that hunt in darkness all too often fears to be alone.

'Men fear death' said Shakespeare's Caesar, 'as children fear to go in the dark.' But not only children fear the dark. Many people suffer feelings of nameless dread as dusk closes round them in some lonely place.

Their apprehension is understandable. How different everything seems when daylight disappears and the darkness becomes full of strange sounds. Things *do* go bump in the night! There are squeaks and grunts and screeches and gurgles and rustles and plops. Stones rattle, shingle slides, bushes move and swish, eyes gleam, ripples spread across the water from unseen swimmers. The creatures of the night are stirring, and at some time or other each of them will make its presence known.

In the blackness, owls hoot and hiss and snore. A vixen screams with scalp-prickling intensity from the dark fellside. In late evening stillness the grunts of feeding badger, or even hedgehog, can be frighteningly loud. A hollow thumping of rabbits on cropped turf; the piercing squeal of a stricken doe as a stoat fastens its teeth in her head; the red stag's bellow, echoing in weird cadence through the silent woods at autumn twilight; the churring of a nightjar . . . Any of these can unnerve a lonely walker overtaken by the night, and send him stumbling even more fearfully homewards.

To anyone who has not experienced it, the coughing of a sheep in the empty mists of dusk or daybreak can be hair-raising, so human does it sound. The sudden appearance of huge out-stretched wings silhouetted against a fading sky, with the heron's harsh 'Kraak' as it glides low towards the margin of its frog-croaking pond, has made many a heart leap with fright, so quietly has the bird approached. Even the cockchafer, the 'shard-borne beetle', whirring like some winged horror in the half-light, can terrify the ignorant.

To those of us who come unprepared into this elemental and seemingly hostile jungle of the dark, these harmless animals are strangers; the noises they make are frightening because they are not understood, and what man does not understand—he fears.

For this reason the owl was long prominent in stories of the supernatural. It is a bird of more than usual interest. For a vertebrate its hunting ability is unique. Probably, no animal has contributed more towards man's feeling of uneasiness at night. Mainly the tawny owl, whose long, wavering call sounds so eerie in the darkness. Tawny owls are particularly vocal late in the summer when the adults defend their territories against their young.

Barn owl, sometimes called white owl, screech owl, church owl, gillihowlet and (by Thomas Bewick) yellow owl. In *A History of British Birds*, 1826, Bewick writes: 'The Yellow Owl is often seen in the most populous towns, frequenting churches, old houses, maltings and other uninhabited buildings, where it continues during the day, and leaves its haunts in the twilight in quest of prey. It has obtained the name of Screech Owl from its cries, repeated at intervals, and rendered loud and frightful from the stillness of the night. During its repose it makes a blowing hissing noise, resembling the snoring of a man.'

Then the barn owl; the 'white' owl that glides ghost-like over the water meadows at dusk with its terrifying screech. The sight of a white shape that floated screaming into the night must have started many a ghost story. As Gilbert White noted: 'White owls often scream horribly as they fly along. I have known a whole village up in arms on such an occasion, imagining the churchyard to be full of goblins and spectres.'

To our forebears, darkness was a symbol of death. It was natural that the mysterious owls, being pre-eminently creatures of the night, should become associated with death and disaster. The belief that owls are birds of ill-omen exists even today.

Undoubtedly, such superstition has its origin in the uncanny ability of some species of owls to hunt not only at night but in utter darkness. This had always seemed to be miraculous. The owl has remarkably good night vision; but no creature, however good its sight, can *see* in utter darkness.

Until recently it was thought that an owl's eyes responded to infra-red radiation from its prey. But this has been disproved. Experiments have shown that in the daytime, on starry nights and in moonlight, owls can certainly hunt by sight; but on very dark, cloudy nights, they locate and pin-point their prey by *sound*—a proclivity made possible by their noiseless flight and asymmetrical ears, owls being the only vertebrates to have such ears.

Of the ghostly silence of an owl's flight, that fine naturalist Charles St John observed: 'If we take the trouble to examine the manner of feeding and the structure of the commonest birds—which we pass over without observation in consequence of their want of rarity—we see that the Providence that has made them has also adapted each in the most perfect manner for acquiring with facility the food on which it is designed to live. The owl, that preys mostly on the quick-eared mouse, has its wings edged with a kind of downy fringe, which makes its flight silent and inaudible in the still evening air. Were its wings formed of the same kind of plumage as those of most other birds, it is so slow a flier that the mouse, warned by the rustling of its approach, would escape long before it could pounce upon it.'

St John was certainly right. But the silence of an owl's wings works two ways: it undoubtedly prevents the mouse from hearing the flight of the owl, but it also enables the owl to hear the rustle of the mouse.

When we think of their exceptional eyesight and sense of hearing, their soundless flight and eerie, terrifying calls, it is not surprising that owls occupy so large a place in country legend. Nor is it surprising, even in these so-called enlightened days, that for so many people the lonely darkness still holds vague feelings of unease.

Wild-life hotel.

But anyone who interests himself in wild animals and can identify their signs and sounds need no longer fear 'things that go bump in the night'. Indeed, the countryside will present a world of fresh interest and fascination, and he will await the coming of dusk with a new-found eagerness.

Alone in the summer twilight, with bats flickering above the treetops and a barn owl hunting silently along the hedge, he may remember those evocative lines of Meredith's:

Lovely are the curves of the White Owl sweeping
Wavy in the dusk lit by one large star.

and watch with delight as the owl swoops low over the shadowy fields, with Venus brilliant in the west.

'The more man has to battle against Nature, the more utilitarian and practical will be his outlook: and it is easily understood why our remote forefathers took more notice of the shriek of a bird which disturbed their hunting, or warned them that they were being hunted, than they took of the tints of its plumage.'

Stephen Potter and Laurens Sargent, *Pedigree: Words from Nature*. (1973)

Mainly from my window

To be out in the countryside at night in bright moonlight has a magic of its own, especially in the silence of snowy winter woods when every sound is magnified and an owl's cry or the distant snapping of a twig seems unnaturally loud. But of all times, daybreak and dusk are the best for watching wild animals. Preferably the early morning, for then the light is growing and one's eyes are less likely to play tricks.

Once from the cottage at sunrise I witnessed an act that few people can have seen. In the meadow not twenty yards away, a fox appeared like a shadow and took a newly-born lamb. Knowing I could never reach the door in time, I hammered on the window and shouted. But I could have saved my breath. The fox looked up at me, cool as you please, the lamb dangling from its mouth; then, quite unhurried, turned and trotted off.

The kestrel, that modern hunter of the motorways, Gerard Manley Hopkins's 'dapple-dawn-drawn Falcon', seems to be aloft within sight of my window at any hour of the day. For several years, kestrels nested in a Scots pine in a little wood up behind the cottage—right next door to a carrion crow. Rather odd, I thought. But both species reared their young successfully.

Close to this pine wood is a little beck that flows down from the Latterbarrow Crag. A local heron used to fish it for eels. And sometimes, after it had finished a meal, the heron flew up and tried to settle in the crow's tree—and out came the crow like a fighter-plane and drove it away. The heron wasn't bothered much. It just flapped off and fished the beck higher up.

Heron footprints are like those of any other wading bird, only very much larger. Underneath several inches of water, etched in the mud of a clear river backwater, or pond, the huge prints seem to be set in glass.

Of the local owls, the barn owl is the first to be seen out and about. In late evening, but before the sun has set, it swoops ghostlike from a ruined barn beside the cottage and glides on silent wings across the moss, hunting mice. Later, at twilight time, the tawny owl's eerie hooting comes wavering across the valley

Left
Ivy berries. They ripen in May and June, and although birds eat them they are poisonous to man.

Below
Rook pellet. One of many found underneath nests in a small rookery. It contains pips and pulp fragments from ivy berries. (Several of the trees were ivy-covered.)

from wooded fellsides. And later still, the tremulous hoo-hoo-hooing of the long-eared owl.

It has been said more than once that Shakespeare was wrong when, in *Love's Labour's Lost*, he wrote of 'the staring owl' singing 'Tu-whit, tu-who, a merry note . . .'. But recently in the early dusk I watched a tawny owl that was sitting on a dead tree in my garden. This bird, which was only a few yards from me, kept up an almost continuous sequence of calls, starting with a 'Ker-wick' and going straight on with 'Hoo-hoo-hoo-hoooo'. And sometimes, after a brief pause, calling 'Hoo-hoo-hoo-hooo' and finishing with a 'Ker-wick', the two utterly contrasting calls being given, so to speak, in the same breath. And to me, that translates perfectly into Shakespeare's 'Tu-who; Tu-whit, tu-who . . .'

I feel sure that the great man heard an owl similar to mine and noted it in that miraculous rag-bag of a memory—later to be captured for ever in his marvellous evocation of a winter's evening.

I have often wondered why Gerard Manley Hopkins ignored the buzzard. He, of all poets, I would have expected to be moved by that haunting, elemental cry. Echo of the wilderness. Mewing spirit of dark woods and hanging cliffs.

At dawn, sometimes, a buzzard swoops low across the valley meadows hunting moles as they move along their surface runs, as they tend to in the early morning. It has the cottage in the middle of its territory and sits on a fence post not thirty yards from the kitchen door. In spring, the rooks come out in force to mob it and drive it away from the rookery—which is just across the lane.

Rookery in pollarded willows. An unlikely site—which has, from necessity, replaced local elms stricken by disease.

Signs of Dutch elm disease: peeling bark and absence of leaves. The disease is caused by a fungus-carrying bark beetle, *Scolytus scolytus*. As disease spreads throughout a tree the bark loses its moisture and (helped sometimes by the stripping of woodpeckers) flakes off in the wind.

Rook's nest in hawthorn. The female (with outstretched wings) is displaying to the male (extreme right) who has just returned from a foraging expedition with his pouch (arrow No. 1), bulging with food — earthworms, leatherjackets and wireworms. The two young rooks, survivors of a clutch of four, are almost fully fledged. The black bristles, which for a year distinguish young rooks from their bare-skinned 'whitefaced' parents, can plainly be seen (arrow No. 2.).

Rook pellet containing vegetable matter and grit (arrowed). Tiny stones and pieces of grit help to grind up food in the bird's gizzard.

Most species of birds eject unwanted items from their meals, including feathers, fur, sand, grass, bones, pieces of shell and other indigestible items. Such 'pellets' vary greatly in size. A robin's is less than one centimetre; whereas a herring gull's may be from four to six centimetres. Pellets both advertise a bird's presence in a particular place and give much information about its diet.

Pellets lying beneath the roost of a long-eared owl.

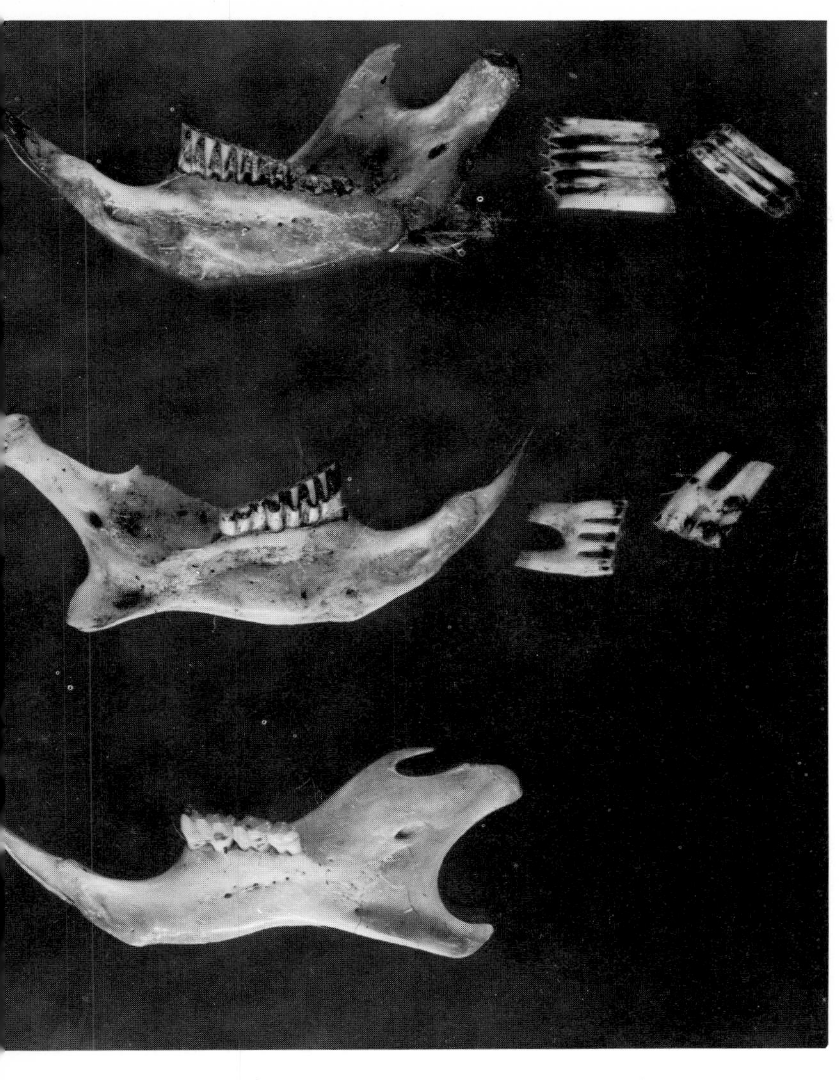

Parts of jawbones and teeth discovered in the pellets of a long-eared owl.

Top to bottom:
Common vole.
Bank vole.
Woodmouse.

Some people dislike having rooks nesting close to the house. But I don't mind. I like rooks. I like watching them. I like the raucous noise they make, especially when the young birds are being fed—a kind of gurgling, jolly sort of sound. Years ago I directed a film about them: *The Riddle of the Rook*. An apt title. There is so much that isn't known about their behaviour. For instance, how they manage to regulate their numbers so efficiently.

What difficult birds they are to film. So crafty. So observant. All the corvids are difficult, but rooks . . . Up there in the tree-tops. So many pairs of eyes, all so sharp. Seeing everything and giving nothing away. At least, very little. Like many other birds, they sometimes divulge what they have been eating by ejecting pellets of indigestible food remains.

Opposite
Little owl, with earthworm. Now a British resident, the little owl was introduced in the latter part of the 19th century. It is widespread in England and Wales, though rare in Scotland and Ireland.

Above
Pellet of little owl. Note the number of beetle wing cases.

Below
Pellet containing matted fur, and a solid red rubber ring presumably from a child's toy. The origin of the pellet is unknown, but it was found in an area frequented by buzzards.

Banks of rosebay willow-herb below my cottage. In the background, Raven Crag.

Wildlife hotel

Above a stretch of sphagnum bogland behind the cottage is Raven Crag. Well named. Ravens have nested there ever since I first knew it. On a seemingly sheer cliff face a white streak of droppings pinpoints the nest site. On still summer evenings I hear the parent birds croaking as they fly overhead across the valley. Even at great range the 'oar'-like wing beats and diamond-shaped tails distinguish them from carrion crows.

Ravens are the aerobatic experts of the avian world. I was an aerobatic pilot myself, years ago. But I give the ravens best any day. I think most pilots would who have seen them at it, as I have seen them. A family group of three or four birds flying in an up-draught above the crag, zooming down in a tight 'V' formation, then swooping up into the blue sky and breaking away like a Prince of Wales's plume of black feathers—before joining up again and showing off their repertoire of stalled-turns, flick half-rolls and tight, flashing aileron corkscrew turns as they hurtle almost vertically earthwards, twisting as they go with a resonant thrum of flight feathers.

Marvellous birds to watch.

In summer the bogland below the crag is white with cotton grass, splashed with yellow marsh bird's-foot trefoil and bright blue clumps of water forget-me-nots. Hidden among the moss is the sundew, the tiny insectivorous plant that feeds on the flesh of ants and midges and mosquitoes, which it traps with the sticky tendrils on its leaves. Seen in huge close-up, trapping and eating an insect, as filmed for our programme *Tender Trap* by Sean Morris and Peter Parks of Oxford Scientific Films, it provides one of nature's more gruesome and horrifying spectacles—and yet, hauntingly beautiful.

In spring and summer the valley is ablaze with wild flowers. Banks of rosebay willow-herb and purple loosestrife grow just below the cottage, and among the bog myrtle in the moss is the spotted marsh orchid. By the duck pond in the water meadow are sneezewort and valerian and ladies smock, and all along the river bank are masses of meadow-sweet, wild thyme, harebells, cranesbill, self heal, speedwell and a profusion of other species, all very common but with such uncommon and endearing names.

Perhaps of all wild flowers, my favourite is the beautiful little betony. The 'sea trout' flower I call it, for it always seems to appear when the main summer run of sea trout comes upstream from the sea. And betonies are still in bloom when the last of the autumn fish are running.

When I go down to fish for sea trout in the summer twilight the air is heavy with the scent of bog myrtle, for me the most evocative smell on earth. More so even than the honeysuckle growing in profusion along the lane that leads to the river.

Overleaf
Morning exercise along the river bank. Always, there are animal 'stories' of the night's events. Signs of birds and mammals—and even fish! A sudden flash of silver, a swirl on the surface, may indicate the arrival of a salmon or sea trout that under cover of darkness only a few hours previously, has forced its way upriver from the sea.

There is no sound of engines or of human voices, only the rustle of a night wind in the woods. Downstream, where the water still glistens in the fading light, a cloud of gnats are dancing. An early bat flickers above the trees. A roe buck barks briefly from the fellside, just audible above the distant murmur of water at the weir. Then the sudden loud splash of a leaping fish shatters the valley stillness and sends ripples racing across the pool to lap the shingle at my feet.

For a moment I stand quite still, transported with a strange sense of timelessness into a world of long ago—when all around me was forest and swamp, and wolf or hyena bayed the moon, and in the deep myrtle-scented dubs the sea trout leaped as they leap today.

And to me it is all sheer magic.

The salmon and sea trout that run upriver in summer are migrants of course, coming in after a miraculous journey from ocean feeding grounds to spawn in the upper reaches of the parent river, or the quiet gravel reaches of some tiny tributary. And it is nature's travellers, such as these, that stimulate my sense of wonder.

Sometimes in February, I catch the wild music of Bewick's swans as they fly past up-valley from the estuary. Having rested on a sand bank at the river mouth after flying from their winter quarters in Ireland, they sweep trumpeting past my cottage on the next stage of their long flight to Siberian breeding grounds.

There is magic, too, in the astonishing journeys of the Africa-wintering warblers, redstarts, whitethroats and other migrants that are nesting in and around the garden, and the wheatears ensconced in their disused rabbit burrows on the fellside. In its own mysterious way, the young cuckoo that is growing fat in a dunnock's nest in the hedge, will soon be flying off to East Africa: How does it find its way? Its parents will have already gone. This young bird will have no one to show it where to go, or how to get there. And yet get there it will.

Swallows seem touched with a special magic. At odds of three to one against, two survivors of the twelve thousand mile flight to and from South Africa are here again, nesting on a ledge in the garden toolshed where they nested last year. And as they sit twittering on the clothes-line that stretches between two damson trees, I picture them on their fantastic flight, flickering across scorching Sahara sands towards some lonely oasis, passing Bedouin camel trains that are heading for the cool water of the same resting place.

The cottage garden with its fruit trees and bushes and ivy-covered dry-stone walls is a breeding place for numbers of small birds and mammals. When we first came to live here it was a wilderness, with pheasants nesting in the long grass. A weasel once tried to rear a nest of young inside the wall just across from my study window—but failed. Puddy saw to that! A weasel would get away with it today, just as the mice do, because Puddy is twenty-one years old this summer and not the hunter he used to be. A young rabbit is living in the wall now. It comes out and feeds a few feet from the window. That would have been impossible when Puddy was younger.

As I write, Puddy is sitting on top of the toolshed trying to snatch the baby swallows from their nest just inside the door—just as he was in my film *Self Portrait of a Happy Man*, and having the same lack of success. The swallows knew exactly where to build their nest.

Although a great hunter, Puddy has always been frightened of strange noises. From my window I once watched him stalking a cock pheasant in the lane. When he had crept almost within pouncing distance of the pheasant, it suddenly looked up and gave a tremendous squawk. Puddy fled—like a streak of white paint!

Years ago he could catch bats. Sat up on the cottage roof behind a chimney pot and grabbed them as they flew past. When he had caught one he would bring it in through the kitchen window to play with—like a mouse—and it would escape, fly up to the ceiling and hang upside-down from a ledge just above my chair. And Puddy would go out and hop on to the roof from the overhanging sycamore and catch another—and this one would do the same. And for a long time I knew nothing about all this. I would come down in the morning and see those bats hanging there, and wonder how on earth they had managed to navigate

Puddy in his heyday.

through a gap in the window narrower than their wingspans. I assumed the bats had been attracted into the kitchen by the moths and midges that earlier had themselves been lured inside by the light. A simple piece of detective work, it seemed to me, until—I came down one morning and found a pair of ears on the floor. . . . Then, suddenly, I understood.

That night I went outside to watch how Puddy did it, and saw him up there in the moonlight like a little white ghost—hiding behind the chimneystack.

Return to the river

Quite suddenly, it seems, the summer has gone. Autumn mists hang over the water meadows and the morning air has a chill in it. The year is dying; the glory of the river fades as the spring migrants gather preparatory to their long flight south.

By November the valley is silent. Masses of leaves drift down under bare trees. Underneath the leaves, late-running salmon and sea trout head steadily upstream towards the spawning redds.

For the bed in which to lay her eggs, the female fish selects a stretch of level gravel in a steady current of clean water. The current is essential, for it not only carries a constant supply of oxygen to the eggs, but helps the hen fish to dig out and, later, to close up the trench in which she lays them.

Many times in my life I have watched the mating ritual of salmon or sea trout. In the clear waters of some upper reach of the main river, or in a feeder stream, every stage can be seen, and it is fascinating.

The hen fish prepares an egg-nest in the gravel by flipping out a shallow trough with her tail. Here she will lay her eggs—seven hundred of them per pound of her body weight. While she digs away, the male fish keeps station nearby, ready to drive off any rival males.

When the trough is finished, and the female is nearly ready to lay, the male goes close up beside her and begins to transmit a violent quivering movement with his body. This stimulates the female, and when the male's quivering reaches a peak she suddenly starts to release eggs.

At once, the male fish ejects milt which drifts back over the eggs like a cloud of smoke, fertilization taking place immediately.

Then, the female covers up the eggs by flipping with her tail from just upstream, so that the current carries the stones into position. This action automatically begins to dig out another trough—in which the whole process is repeated. A pair of fish will stay together for several days, during which the female digs perhaps three or four troughs, laying eggs in all of them.

Two points are worth noting.

First. During the latter stages of her preparation of the redd, the female presses down into the trough with her anal fin. This, seemingly, is to test the depth. At any rate, it informs an observer that she is nearly ready to lay.

Secondly. During the orgasm the mouths of both fish gape wide open—a clear signal that laying has begun.

After spawning, most of the males and many of the females die. But each egg lies secure under its canopy of gravel. And gradually a little fish forms inside the egg, deriving oxygen from the water that flows between the stones.

After a period of two to four months, this embryo hatches out as an alevin—a tiny, translucent, pinkish creature with an umbilical sac hanging down below its throat. The time taken for the egg to hatch depends on the water temperature; incubation being retarded until the water is warm enough to contain a supply of food. The yolk-sac attached to the alevin is a further insurance against starvation, for it holds upwards of a month's rations.

During the alevin stage our little fish lives on the contents of its yolk-sac, and when these are exhausted the alevin becomes a fry.

It is now forced to fend for itself and, in company with other fry, hunts actively for food, gradually acquiring a form of camouflage in the shape of dark 'finger-marks' along its sides. At this stage it becomes known as a 'parr' and in looks is very similar to a small brown trout.

Trout alevin showing yolk-sac.

Overleaf
Heading upstream, this fish leaps a waterfall as it makes its way towards the spawning redds. From the shape of the head and the tail it can be identified as a male sea trout.

Although the trailing edge of a very small sea trout's tail is forked, this fork becomes less and less pronounced as the fish increases in weight. At 4 to 5 lb, the tail becomes square. Upwards of about 8 lb the tail is convex. A salmon's tail is invariably concave.

Not without good reason the Romans dubbed salmon and sea trout *Salmo*—which means: 'The Leaper'. But for centuries, the question of how salmo managed to perform its leap was the matter of some controversy. One school of thought was of the opinion that a salmon 'threw itself over waterfalls . . . by taking her tail in her mouth and bending her backbone downwards, till the letting it go all at once gives her strength enough to throw herself over the cataract at a leap.'

But a more astute observer wrote: 'I very much question the truth of this . . . for neither does the salmon's mouth, which is small and weak, and the tail, which is large and slippery, allow the thing in itself. But that they *will* leap out of the water is certain. I have seen a very large salmon leap five or six feet high.'

He was right. The leaping of salmon has often been exaggerated, but the truth is sufficiently dramatic. A leap of eleven feet has been measured over a perpendicular waterfall.

A dream of clear water. An unpolluted Cumbrian stream, with sea trout fresh-run from the tide.

For two or three years it remains in the river as a parr, feeding on tiny crustaceans and insect larvae. It hides underneath stones in the winter, but feeds during spells of warm weather.

Eventually, the little fish develops a silver coat, becomes known as a 'smolt', and goes to sea. And it goes out to sea in order to grow. This migration is imperative. Few rivers contain a food supply that could support a large population of fully grown salmon and sea trout. Only by going to sea can these fish find sufficient food.

Exactly where young sea trout go is unknown. But it is to

feeding grounds off the coast of Greenland that many salmon go, and where, during recent years, so many thousands of tons have been netted that the species was brought almost to the point of extinction. The fishing off Greenland has been reduced; but on their return to the British Isles, large numbers of fish are being intercepted off the west of Ireland where netting is unrestricted. And far too many of the survivors are being trapped illegally off our own coast.

Some salmon return to fresh water after only a year at sea. Some after two years or more. But at whatever age, return they must, for they cannot spawn in salt water. And astonishing though it seems, the tagging of smolts has shown that most fish come back to the very same rivers they grew up in.

That salmon can find their way back from the distant ocean to the coast is a staggering feat of navigation, and nobody knows how they do it. But a marvellous piece of detective work by two American scientists has proved that once a salmon *reaches* the coast, it can find its own river by sense of smell; by the particular *odour* of that river—the river it left when it went to sea as a smolt. And that, to me, is a miracle in itself.

But if, like so many stories of migration, the salmon's life-cycle seems miraculous, what am I to say of *Anguilla anguilla*, the European eel?

When, in early May, I watch an elver wriggling up the little river below my cottage, I am reminded that there is an example of endurance second to none; that this tiny, thread-like creature, battling on against unbelievable odds, really is the most astonishing animal on earth.

I am recounting the eel's story in some detail because it contains a wonderful piece of scientific nature detective work, and illustrates how little we know about this exceedingly common fish—the first, by the way, to achieve the dignity of a name.

The very history of the eel is touched with wonder. To the ancient Greeks it was the King of fish and, highest of all honours: the Helen of the feast.

The Boeotians thought it worthy of sacrificial offering to the gods.

To the Egyptians, it *was* a god.

Sybaritic eel-catchers escaped their taxes.

By the credulous it was considered a prophylactic, a panacea, a tonic for the voice. The smoke of burnt eel eased the pangs of childbirth, whereas the fragrance of cooked eel restored the sense of smell to a dead man.

The Athenians thought it the greatest of all delicacies. As one of them lamented: 'When you are dead, you cannot then eat eels!'

But for all these panegyrics, no one knew where eels came from or how they bred. It was very mysterious. The supply of eels seemed inexhaustible—and yet, no eel had ever been observed to spawn.

This was hardly surprising; since, as we know today, the European eel does not originate in European waters. Together with its close relative the American eel, it breeds thousands of miles away in the depths of the Sargasso Sea outside the Gulf of Mexico.

From the Sargasso, the eel larva starts off in the springtime as a tiny leaf-shaped creature called a *leptocephalus*, which drifts north-eastwards with the Gulf Stream. Three years later it turns up on our shores as an elver, or miniature eel: a slender creature about three inches long.

Practically no stretch of fresh water is without its complement of eels; and each year a new stock arrives from the Atlantic. During April and May, the elvers make their way upriver towards their various destinations. From then on, an eel lives in fresh water—until, years later, on some wild autumn night, it returns to the sea to spawn.

That, very briefly, is the eel's life-story. But until the present century little of this was known. There is only one species of European eel, but the change of shape and coloration that takes place shortly before its return to sea—notably, the change from yellow belly to silver belly—led early observers to suppose that there were several different species; and each of these species was thought to breed in a different way.

The riddle of the eel's breeding-cycle became the subject of continuous debate. Aristotle records the first important observation;

'Some fish leave the sea to go to the pools and rivers. The eel on the contrary, leaves them to go down to the sea.'

He was right. But as eels had never been found with ova or milt inside them, and were seemingly lacking in generative organs, his imagination took over when it came to explaining how eels bred.

'They form spontaneously,' he suggested. 'In mud.'

Oppian thought that little eels came from eel slime.

> 'Strange the formation of the eely race.
> That knows no sex, yet loves the close embrace.
> Their folded lengths they round each other twine,
> Twist amorous knots, and slimy bodies join;
> Till the close strife brings off a frothy juice,
> The seed that must the wriggling kind produce.
> That genial bed impregnates all the heap.
> And little eelets soon being to creep.'

In Pliny's opinion, eels were sexless—being neither male nor female. 'They have no other mode of procreation than by rubbing themselves against rocks—and their scrapings come to life.'

Other thinkers attributed the birth of eels to the dew of May mornings (the month, note, when elvers are usually running upstream); the hair of horses; the gills of fishes, and to various forms of 'Spontaneous Generation . . . by the sun's heat, out of the putrefaction of the earth.'

Gradually, as the centuries went by, a little light began to dawn. The Reverend W. Richardson, writing in 1793, suggested that eels descending a river went down to the sea to breed. Of the Cumbrian Derwent, he remarks:

'The young eels come up the river in April, in size about the thickness of a common knitting needle.'

This was a very accurate piece of observation. But naturalists in Britain were lagging far behind those on the Continent. That adult eels disappeared into the sea, and young eels came out of it, had been noted in Italy a century earlier. In this country, even as late as 1862, in the *Origin of the Silver Eel*—a work based on 'observations extending over sixty years'—a Mr D. Cairncross blithely asserts:

'The progenitor of the silver eel is a small beetle. Of this, I feel fully satisfied in my own mind, from a rigid and extensive comparison of its structure and habits with those of other insects.'

The first recorded capture of an immature eel, before it had reached the elver stage, was made by a German scientist in the Straits of Messina in 1856. But not realizing that the strange, leaf-like little creature was, in fact, a young eel, he named it *Leptocephalus brevirostris*—or, Short-snouted thinhead.

That this *Leptocephalus brevirostris* was indeed the larva of the eel, was discovered by two Italian scientists in 1896. But it was not until some years later that the eel's spawning ground in the Sargasso Sea was located. This was a classic piece of detective work by a young Danish marine biologist, Johannes Schmidt, who published his findings in 1921.

By means of intensive netting with fine-meshed nets right across the North Atlantic, and examining haul after haul, Schmidt was able to draw up a chart showing that the *leptocephali* he caught became progressively smaller and smaller as he approached the Sargasso Sea. There, he found the smallest larvae of all.

It was, he inferred, from their Sargasso birth-place, that the tiny eels set off as *leptocephali* on their three-year eastward drift, during the latter stages of which occurs the metamorphosis from *leptocephalus* to elver.

Having reached our shores in the late spring and early summer, the millions of little eels that have survived this incredible journey start to fight their way up towards the freshwater destinations where they will spend most of their lives. As the Polish biologist, Opuszynski, has observed: nothing, it seems, can stop them.

'Their urge to undertake this journey is unconquerable. They are not deterred by any obstacle such as sluice or waterfall. They have been seen mounting a vertical wall. Even the bodies of the elvers which die in doing so and adhere to the wall, serve as a kind of rung for succeeding elvers.'

But, most mysterious of all, what strange impulse drives one elver to wriggle on and on upstream, mile after mile, to some distant tarn high among the hills, while another stays where it is in the estuary, or lower reaches of the river, or in some local ditch? Is the little eel's destination plotted from birth—an inheritance from one of its parents? Nobody knows.

From data recorded by Dr Winifred Frost at the Freshwater Biological Association, Windermere, the male eel remains in fresh water for seven to nine years; the female for ten to twelve years—although in one exceptional case, a stay of nineteen years has been recorded. Almost invariably, as in the case of pike, the mature female eel is larger than the male. Female silver eels—that is, eels on the point of seaward migration—average about 1 lb, whereas the males average 3–4 oz, and very rarely exceed 18 inches in length. Large eels are almost always female.

After spending their alloted time in fresh water, the eels return to sea. In preparation for this migration, the snouts of both sexes become more pointed; the eyes enlarge; the back becomes black; the belly changes from yellow to silver, and the gut degenerates—so that, like many other migrating fishes—salmon and lamprey, for instance—the eel does little or no feeding during its arduous journey home.

These changes take about six months to complete, the journey seawards starting during late summer and autumn. But migrating eels seldom move downriver if there is a glimmer of light. They are creatures of darkness, tending to travel only on wild, cloudy nights, and usually when the river is rising. As Buckland observed:

'Not in calm, clear nights: the darker and stormier the night, the greater the exodus.'

It is on some wild autumn night, then, that the European eels start to nose their way seawards—seemingly towards their distant Sargasso spawning grounds.

At least—they *do* go to sea. That much is certain.

But little more is known. As Leon Bertin says in his biological study of the eel:

'The eels virtually disappear once they have reached the sea, and we are almost completely in the dark concerning the tremendous journey of many thousands of miles which they

What fisshe is slipperer than an ele?
Ffor whan thow hym grippist and wenest wele
Too haue hym siker right as the list,
Than faylist thou off hym, he is owte of thy fyst.'

Piers of Fulham, *c.* 1400

Mature European eel. After a fantastic journey as a *leptocephalus* from the Sargasso Sea, metamorphosing into an elver, or baby eel, and spending years of its life in fresh water, the adult returns to the ocean to spawn. And as it swims downstream in the autumn darkness, it passes the incoming salmon swimming upstream from the sea to spawn in the river. Both species moved by the same desire—to procreate. The salmon having just come thousands of miles, the eel with thousands of miles to go.

Fish that pass in the night.

accomplish in their passage to the Sargasso Sea.'

But do they, in fact, ever reach the Sargasso?

In this respect, even today, the story of the eel is incomplete. A theory has been published suggesting that the European eel is not equipped for such an exhausting journey; that European stocks derive from the *American* eel—which provides larvae for both American and European waters.

The principal difference between American and European eels is an average difference in the number of vertebrae; about 115 for the European eel; 107 for the American.

It is claimed that a difference in water temperature at which eggs develop and hatch can result in a difference in the number of vertebrae. If this is so, eggs that hatch in the westerly waters of the Sargasso may drift off to the north-west and become American eels, while those that hatch in the eastern waters drift north-east and become European eels.

On the face of it the theory seems unlikely. But no one has disproved it. Indeed, no one has even found an adult eel on the supposed Sargasso breeding grounds, and at the present time the mystery surrounding this vital part of the eel's life-cycle remains to be solved.

What an opportunity for some nature detective of the future.

Tailpiece

A little while ago one baby eel actually finished up in the cottage. Our water supply comes from a beck fed by a spring on the fellside. And this particular elver had wriggled from the river all the way up the little beck, and then all the way down the feed-pipe into the water-tank in the attic.

Just think of it.

This tiny creature starts off thousands of miles away in the depths of the Sargasso Sea. For three years, escaping a host of predators, it drifts across the Atlantic with the help of the Gulf Stream until, of all places along the coasts of Europe, it reaches Ravenglass. And then, for reasons known only to nature, this little miracle of survival finds its way up this particular river into the particular beck that provides our water supply.

It has lived through countless dangers during one of the most astonishing journeys in the world—and it ends up in my tea kettle!

That's ridiculous.

Picture credits

The author and the publishers acknowledge permission to reproduce the photographs in this book from the following:

Jeffery Boswall, 120, 121
J. B. & S. Bottomley, 101
Fred Buller, 110, 111, 168, 169
Dr G. Brouwer, 118, 172
Jan van de Kam, 45, 93, 98, 100, 102, 103, 104, 105, 108
John Clegg, 162, 164, 245
Geoffrey Kinns, 160, 172, 190
Hans Kruuk, 24, 36, 51, 52, 56, 57, 66, 170
Lea MacNally, 171, 182, 183
Arthur Oglesby, 134, 138, 139, 146, 150, 152, 166, 167, 178, 181, 183, 184, 185, 186, 187, 188–9, 189, 190, 191, 192, 195, 196, 197, 198, 200, 201, 202, 203, 204, 205, 206, 208, 209, 210, 211, 212, 214, 215, 216, 217, 218, 219, 220, 221, 229, 237, 238, 240, 246, 248
F. van Ommen, 170
Frank O'Neill, 51, 112, 148
Tom Rawling, 17, 25, 34, 58, 95, 96, 97, 104, 109, 126, 137, 147, 152, 155–6, 158–9, 160, 186
Bernard Roughton, 231
R.S.P.B., 223
Dr B. J. Spencer, 18, 78, 79, 86
Dr R. Summers, 114, 115, 116, 117, 118
Fred J. Taylor, 253
Prof. Niko Tinbergen, endpapers, 8, 14, 15, 18, 19, 20, 22, 23, 25, 26, 27, 28, 29, 30, 31, 32, 33, 35, 37, 38, 41, 42, 43, 44, 47, 48, 50, 55, 57, 58, 59, 60, 61, 62, 63, 65, 67, 68, 69, 73, 74, 75, 76, 80, 81, 82, 83, 84, 87, 88, 89, 90, 91, 106, 113, 119, 122, 124, 125, 126, 127, 128, 131, 142, 146, 156, 173, 231, 234, 235
M. C. Wilkes, 2, 39, 40, 49, 53, 72, 136, 137, 140, 141, 142, 143, 144, 145, 147, 149, 151, 152, 153, 154, 161, 174, 175, 176, 193, 199, 224, 227, 232, 233, 234, 236, 237
Dermot Wilson, 165